Plant & Animal Structure & Function

Biology Lab Manual

First Edition

Jonathan Christie

Chemeketa Press | Salem, Oregon

Plant & Animal Structure & Function: Biology Lab Manual

ISBN-13: 978-1-955499-33-0

Chemeketa Press
Chemeketa Community College
4000 Lancaster Dr NE
Salem, Oregon 97305
collegepress@chemeketa.edu
chemeketapress.org

Cover design by Ronald Cox
Interior design by Ronald Cox and Abbey Gaterud

A full list of credits appears on page 163 and constitutes an extension of this copyright page.

Printed in the United States of America.

Land Acknowledgment
Chemeketa Press is located on the land of the Kalapuya, who today are represented by the Confederated Tribes of the Grand Ronde and the Confederated Tribes of the Siletz Indians, whose relationship with this land continues to this day. We offer gratitude for the land itself, for those who have stewarded it for generations, and for the opportunity to study, learn, work, and be in community on this land. We acknowledge that our College's history, like many others, is fundamentally tied to the first colonial developments in the Willamette Valley in Oregon. Finally, we respectfully acknowledge and honor past, present, and future Indigenous students of Chemeketa Community College.

Contents

Lab and Safety Regulations

Personal Behaviors

1. Eating and drinking are prohibited in the labs.
2. Wash your hands before leaving the lab.
3. Wash the lab counter & related work areas with disinfectant before and after the lab.
4. Stow your personal items in the cubbies or under your station.
5. Keep the lab counters and work areas uncluttered.
6. Clean-up is your responsibility:
    a. Return cleaned items back to lab kits.
    b. Wash slides and return to the original location (unless told otherwise). Cover slips may be thrown away.
    c. Wash glassware and return to original location (unless told otherwise).
    d. Ensure any lab waste is disposed of in the approved container(s).
    e. Ensure sinks and counters are as clean as when you arrived.
7. Clothing must be appropriate for the lab to be performed.
8. Read the lab in advance so you're aware of proper safety protocols.
9. If you have questions about safety, ask before you do.

Safety Protocols

1. Know where the lab safety equipment is located–fire extinguisher, first aid, eye wash, etc.
2. Wear personal safety equipment (goggles, gloves, aprons) as indicated in the lab instructions or by your instructor.
3. Handle all chemicals and biologicals, including stains, below eye level.
4. Do not assume something can just go down the sink or in the trash. Dispose of wastes in appropriate waste containers.
5. If there is a chemical spill:
    a. Get your instructor.
    b. If the spill is on the floor or counter, keep away from it.
    c. If the spill is on you or your clothes, rinse the area immediately with running water.
    d. If you splash chemicals in your eyes, go immediately to the eye wash station, turn on cold water, remove the red caps and lean down so that the water bubbles into your eyes.
6. In case of injury:
    a. Get your instructor.
    b. If the instructor is not available, call campus safety (x5023) on the lab's phone.
7. If the injury appears severe, also call 911
8. In case of fire:
    a. If your clothes are on fire, yell "FIRE" and roll on the floor or use a coat or fire blanket to smother it.
    b. If chemicals or lab equipment is on fire, call out "FIRE" and evacuate the room.
9. If the fire alarm goes off, take your essential items and leave the room. Head to the evacuation area as told by your instructor. Stay with your class and wait for instructions. Do not re-enter the building until an all-clear is given.

Name: ______________________ Lab Time: ____________ Due: ______________

1 Animal Body Organization Pre-Lab

1. How are a cell, tissue, and an organ different?

2. What are the three types of connective tissue you'll discuss in today's lab?

3. What are three things you'll identify on the skull in Activity 1.3?

4. Activity 1.3 references quadruped and biped. What do these terms mean? What term applies to a cow or a deer? What about a human?

5. Look up the word sagittal. What does that term mean? Give an example of something in the lab to which this word applies.

Name: ______________________ Lab Time: ____________ Due: ______________

1

1 Animal Body Organization

All organisms you're used to seeing in your daily life have organization. That is, they have layers, components, and functions that are consistent across species and time. For example, animals have a way to maintain support against gravity (what we call a skeletal system). Plants have a way to transport water from the roots to the leaves (what we call a vascular system). Seeing these common organizational elements is one way to visualize how life reflects an interaction between homeostasis and evolution.

Learning Outcomes

- ❑ Identify animal body systems and their roles in homeostasis
- ❑ Use anatomical observations to infer lifestyle and/or behavior

Background

Large structures (such as an entire organism) are often made up of smaller structures working together–just like at Chemeketa, where one college is made of many departments. This **hierarchy of the body** gives the relationship between body parts and their function(s) in relation to each other and to the organism as a whole.

The hierarchy we're going to use in this lab is:
organism → body systems → organs → tissues → cells

Q1. What term represents the largest (most easily visible) level of the hierarchy?

Q2. What level(s) in the hierarchy are considered "alive?" How do you know?

Q3. Do you think all organisms have the same body systems? If so, do you think they have the same organs?

Q4. Body systems are often labeled after their function. What do you think is the function for these representative systems?

- ❑ skeletal system:
- ❑ digestive system:
- ❑ reproductive system:

Activity 1.1. Animal Tissues

Tissues are a collection of cells working together to perform a task. Animals contain four major tissues: **epithelial tissue**, **connective tissue**, **muscle tissue**, and **nervous tissue**. Each one has unique characteristics and roles in the body. For this lab, we'll focus on epithelial and connective. We'll talk about the other two in later labs.

Epithelial Tissue

Examine Figure 1.1 (or one from your text) showing different types of epithelial tissues.

Figure 1.1 Different types of epithelial tissues.

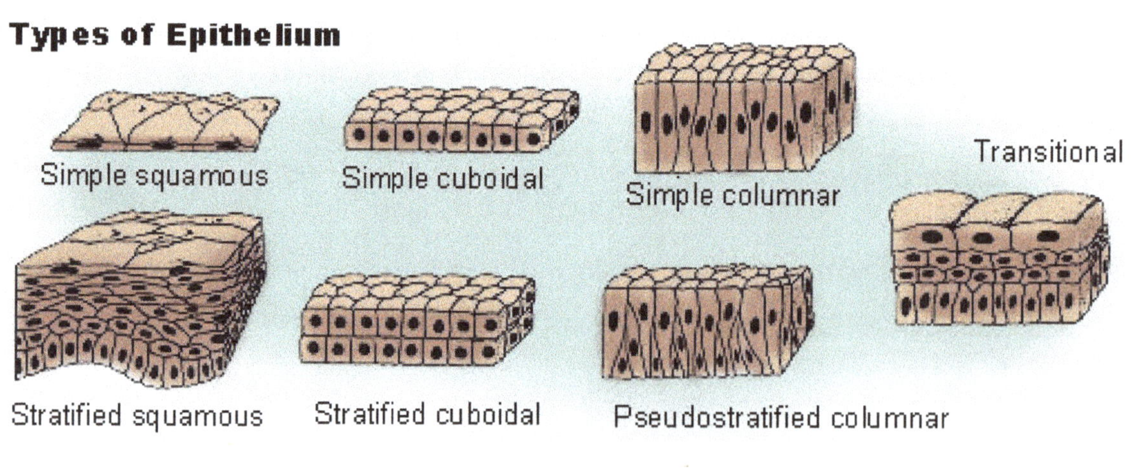

Q5. Circle a single cell in Figure 1.1. What is the center of the cell called? Draw an arrow to it.

Q6. What structural similarities do you see in the different epithelia shown above? In other words, what do they have in common? What about differences?

Q7. What does "epi-" mean? From that prefix, what can you conclude about the location of this tissue in an organ?

Q8. Which of the epithelia in the diagram would be the most protective of an organ? Why?

Connective Tissue

Many types of connective tissue exist, with different appearances. The characteristics of each are beyond the scope of this lab. Instead, we'll focus on three that you'll use later in this course: bone, blood, and fat (adipose). Use the internet to briefly describe the organization of each of the following types of connective tissue:

- ❑ Bone:
- ❑ Blood:
- ❑ Fat (adipose):

Q9. What is a direct relationship between bone and blood?

1

Q10. How do all three of the connective tissues above physically "connect" the body?

Activity 1.2. What Skin Tells Us

Procedure A. Examining Skins

Instructions

1. Obtain the laminated photos of seal surgery.
2. Sketch what you see in Figure 1.2.
3. Label: epithelial tissue and a possible connective tissue in Figure 1.2.

Figure 1.2. Sketch and label the tissues present in the seal surgery photos.

Q11. What do you notice about the thickness of layer #2 compared to layer #1 in the photo? Why might this be?

Q12. Layer #2 is called blubber. **Blubber** is an insulator and energy reserve for marine mammals. Based on this, which specific tissue do you think blubber is made of?

Q13. Make a hypothesis about the thickness of blubber in relation to seawater temperature. What homeostasis relationship might there be?

Examine the graph in Figure 1.3, which shows the average blubber thickness over several months.

Figure 1.3. The average of blubber thickness over several months.

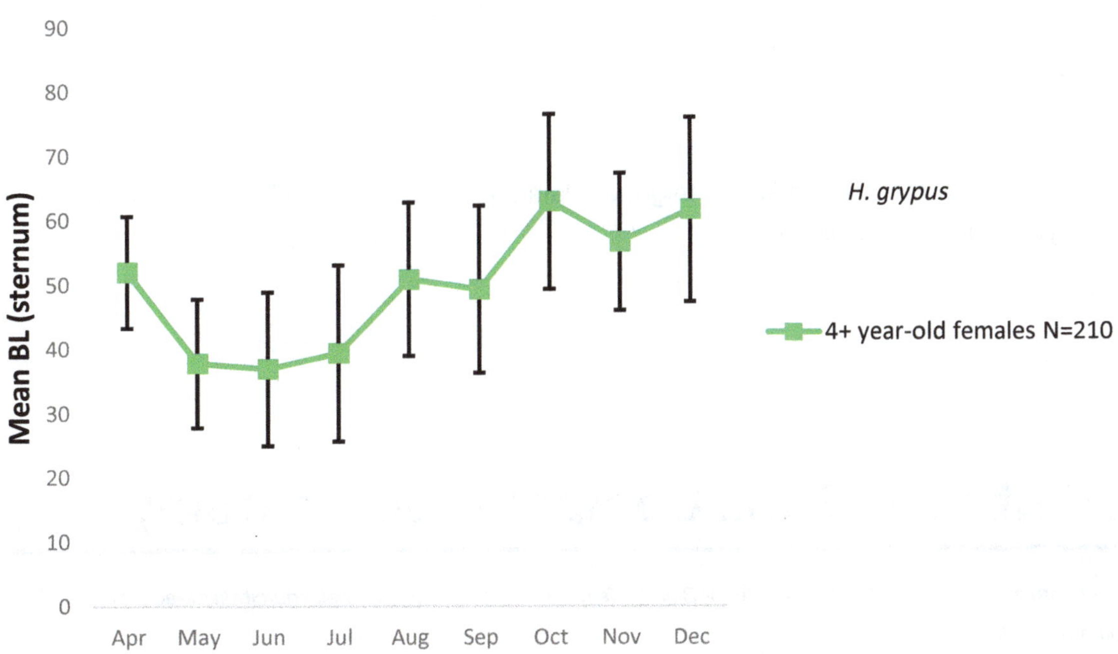

Q14. What conclusion can you make from the graph in Figure 1.3? Does this graph support your hypothesis? Why or why not? Relate your answer to homeostasis.

Figure 1.4 is another graph from the same article.

Figure 1.4. The average of blubber thickness over several years.

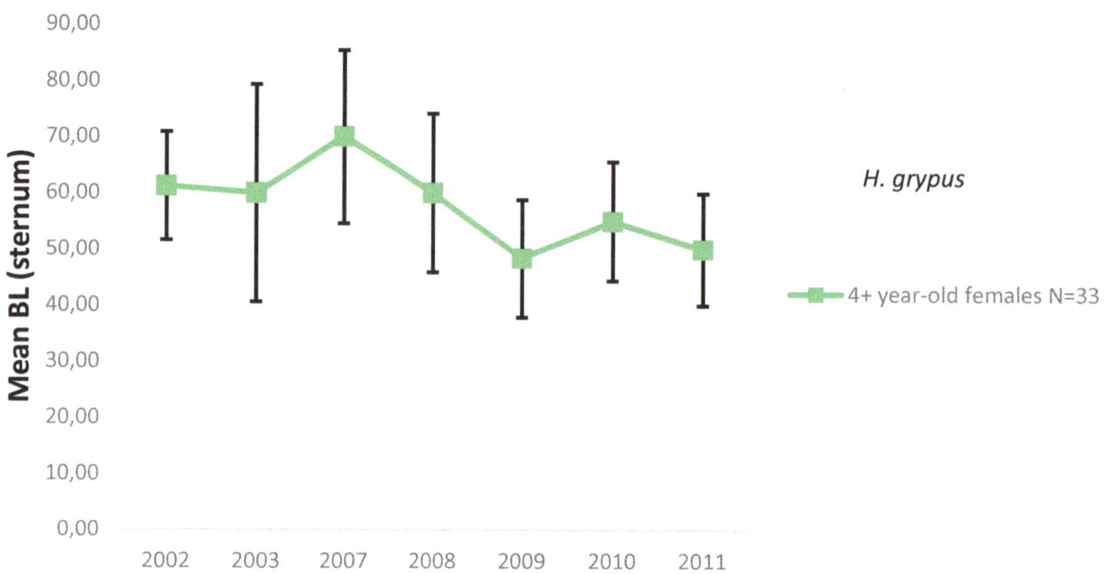

Q15. What environmental conditions might explain the trend shown in Figure 1.4, in light of your earlier conclusions?

Activity 1.3. Skull Comparative Anatomy

(adapted from a similar lab at Cornell Institute for Biology Teachers = https://blogs.cornell.edu/cibt/labs-activities/labs/comparative-skulls/)

Learning animal anatomy isn't just an academic pursuit. You can use anatomy to answer questions about an organism's lifestyle and evolution by examining the physical clues in its tissues and organs. For this part of the lab, we'll be looking at animal skulls and using key characteristics to answer questions such as:

- ❑ the approximate length (including the tail) of the original specimen
- ❑ whether the specimen was a predator or a prey species
- ❑ whether the specimen was an herbivore, omnivore, or carnivore
- ❑ how well it could see or smell
- ❑ whether it was a quadruped or a biped
- ❑ how much muscle mass moved its jaws

Procedure A. Examining Skulls

Instructions

1. Get the skulls provided by your instructor. Each skull should have a letter/color mark to identify it. You will use that identification in your worksheet answers.
2. Measure each of the following and record them in Table 1.1.
   A description of how to do each measurement is provided on the laminated sheets on your lab bench.
3. Length from the *foramen magnum* to the end of the snout (in cm). Record twice and average.
4. Eye angle to skull (in degrees).
5. Major teeth types. (*incisors, canines, premolars, molars, carnassial*).
6. Presence of *sagittal crest* (*developed, moderate, none*).
7. Complexity of *turbinates* inside the nose. (*simple, moderate, complex*).
8. Placement of *foramen magnum* in relation to spinal cord. (*bottom of skull, back of skull, in between*).

Table 1.1. Observations and Measurements of Skulls

Skull	snout length (average, in cm)	eye angle (degrees)	teeth types (I, C, P, M)	sagittal crest?	turbinates?	f.m. & spine relationship
A						
B						
C						
D						
E						
F						
G						

The sagittal crest is important to determine jaw muscle strength. Does the presence of the sagittal crest indicate comparatively strong or weak jaw muscles? Why? How does this relate to the function of the skeletal system?

Q16. Use your data from Table 1.1 to draw conclusions about each skull specimen and enter them in Table 1.2. You will have to infer some of these from multiple data points in your chart, along with general observations.

Table 1.2. Skull Summary

Skull	Total body length in cm and feet *see below*	herbivore, carnivore, or omnivore?	biped or quadruped?	Vision acuity & smell acuity? **see below*	What animal do you think this is?
A					
B					
C					
D					
E					
F					
G					

***** total body length = 5 * average snout length in cm; to convert to feet divide by 30.48 (30.48 cm = 1 foot)***
***** Vision acuity: Good binocular vision, good peripheral vision, or nocturnal.***
***** Smell acuity: Developed sense of smell, moderate sense of smell, poor sense of smell.***

Q17. Examine Table 1.2. Are there any boxes that seem erroneous or don't match the pattern you think? Which ones? Why might it be odd?

Q18. Sometimes animals that we typically think of as four-legged can actually stand on two legs. Will this make a difference in any of your measurements or conclusions? Why or why not?

Q19. Looking at your data on eye angles, what benefit might having a low angle have to that species? What about a high angle?

Q20. Analyze your data regarding smell. Which other factors show a relationship with turbinate complexity? Why might that be?

Q21. Look at the bird skeleton at the bottom left of the display case outside room 8/225 and note the bony keel. You can also see it highlighted in Figure 1.5. Based upon your understanding of the sagittal crest in your skulls, what might be the function of this "crest" of bone? What muscles likely attach here? How does this show a relationship between bone structure and muscle function?

1

Figure 1.5. The bony keel of a bird skeleton.

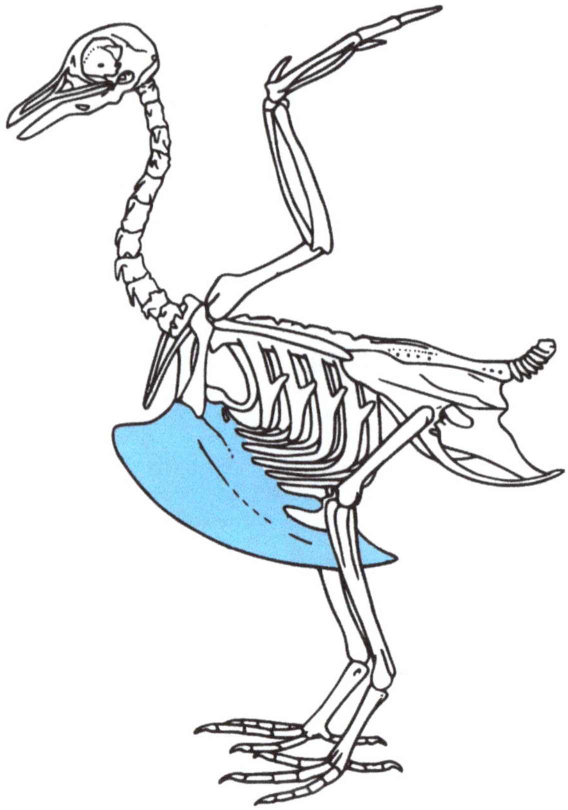

Name: ____________________ Lab Time: ____________ Due: ______________

1

1 Applying What You've Learned

1. Regarding the skulls you looked at, where do they rank in the organism hierarchy? In other words, is the skull a system, an organ, a tissue, or a cell? Justify your choice.

2. Figure 1.6 is an X-ray of a frog. Look at the *urostyle*.

   - ❑ What do you think is the purpose of this structure?

   - ❑ Do you think all amphibians (such as salamanders) have this structure? Why or why not?

Figure 1.6. X-ray of a frog.

1

3. Some skulls show a structure called *auditory bullae*. These are hollow chambers of the skull containing the organs of hearing. What could you conclude if your skull had noticeably large auditory bullae? Why? Give an example of an animal that might fit this description?

Name: ______________________ Lab Time: ____________ Due: ______________

2 Homeostasis Pre-Lab

1. What are the three parts of a homeostasis feedback loop?

2. What are the two types of feedback seen in homeostasis? What is the difference between them?

3. What is a characteristic of capillaries mentioned in Activity 2.2?

2

4. What is one function of the stoma mentioned in Activity 2.3?

5. What are two safety requirements for Activity 2.3?

Name: ______________________ Lab Time: ____________ Due: ______________

2

2 Homeostasis

Homeostasis is a fundamental characteristic of all life. Basically, homeostasis is a process by which organisms respond to internal changes. In other words, homeostasis is a way for organisms to regulate their internal environment in response to an ever-changing world.

Learning Outcomes

- ❑ Explain and diagram a negative feedback loop.
- ❑ Analyze experimental data illustrating possible homeostatic processes.
- ❑ Give examples of homeostasis related to an organism's environment.
- ❑ Compare and contrast evolutionary differences in homeostasis between plants and animals.

Activity 2.1. Homeostasis and Feedback Loops

Homeostasis is maintained by a complex set of physiological processes. To do that, organisms must have a way to coordinate actions between various body parts to achieve homeostasis. This is commonly called a **feedback loop**. It's called a loop because often the end action "loops back" and affects the beginning action in some way. It's called "feedback" because the body adjusts its response based on the loop. Figure 2.1 illustrates this idea.

Figure 2.1. A visual representation of a feedback loop.

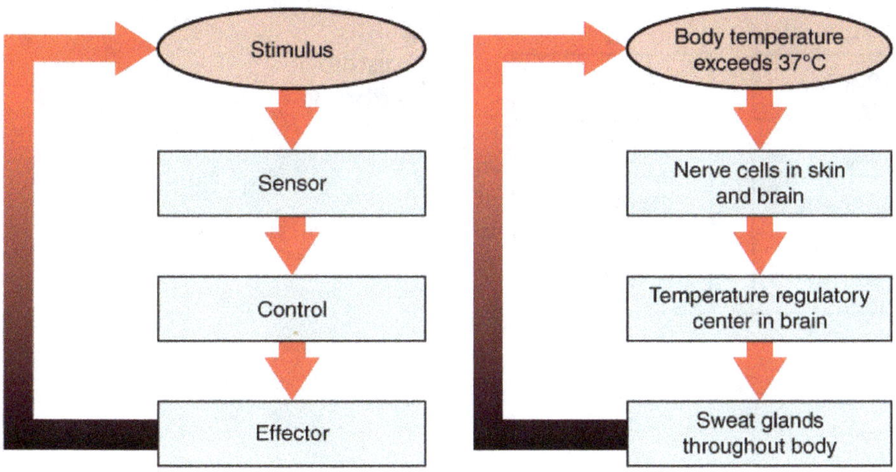

(a) Negative feedback loop (b) Body temperature regulation

Q1. Using the diagram in Figure 2.1, what do you think is the function of a sensor?

Q2. What body system or organs do you think normally perform the sensor's job?

Q3. What are some examples of a "stimulus" that might apply to homeostasis in the human body?

Q4. What is one possible effector in the human body?

Homeostasis is divided between *positive-feedback* and *negative-feedback*. **Positive-feedback** is a response by the body that increases (amplifies) the stimulus. **Negative-feedback** is a response by the body that decreases (reduces) the stimulus. Figure 2.2 illustrates this difference.

Figure 2.2. An example of positive and negative feedback.

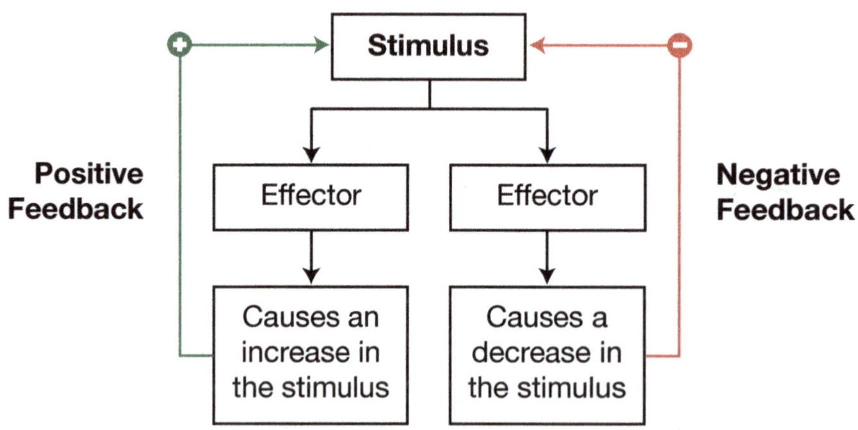

Q5. Using Figure 2.2, predict what action an effector would take if:

- ❑ The stimulus was an increase in blood sugar and the feedback was positive.
- ❑ The stimulus was an increase in blood sugar and the feedback was negative.

Procedure A. Homeostasis Roles

(adapted from: https://thetrendyscienceteacher.com)

You will perform a quick group activity representing the roles (components) involved in homeostasis in the body. The set-up is shown in Figure 2.3.

Figure 2.3. The set-up for Procedure A group activity.

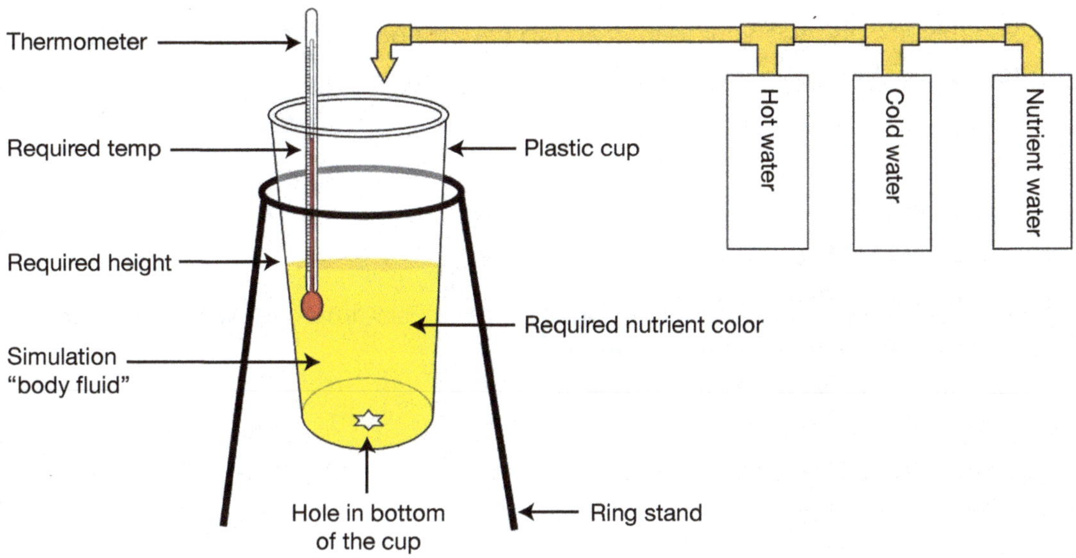

Your team is responsible for keeping the "body fluid" at a consistent temperature, height, and nutrient color for 5 minutes. Your teacher will be the timekeeper.

Before you begin, you must divide your team's tasks. Determine the function of these "jobs" in Table 2.1 and decide who in your group will do each one.

Table 2.1. Divide Your Team's Tasks

	What will you do in this experiment?	Who in your team will do this?
sensor(s)		
controller		
effector(s)		

2

Q6. For this experiment, you can talk and listen to each other. In real life, what are ways organs communicate with each other in the body?

Instructions

1. Set up the experiment as in the diagram on the laminated instruction sheet in your kit.
2. Ensure your team members know their assigned roles and are ready.
3. When you're ready, the teacher will start the timer.
4. Use your homeostasis roles to keep the three requirements (temperature, height, color) the same for five minutes.
5. When the time is up, clean up your area.

Q7. In doing this activity, was your team trying to perform positive-feedback or negative-feedback? Explain.

Q8. Did your success improve over time? Does homeostasis "improve" with time over the life of an organism? Why or why not?

Q9. What body system was represented by these parts of the experiment:

- ❑ The hole in the bottom of the cup?
- ❑ The addition of water and nutrients to the cup?
- ❑ The thermometer?

Q10. This activity had three variables to control–temperature, color (fluid nutrient level), and height (fluid volume). What is an additional variable your body controls regarding your body fluids, other than these three?

Q11. How do you think increasing the number of variables being controlled affects the demand for homeostasis in an organism? Given that, explain why many "simple" organisms live in an aquatic environment rather than on land.

2

Activity 2.2. Skin Temperature Recovery

The ability of your skin to recover from temperature exposure depends, in part, on the blood vessels in your skin. Blood vessels called capillaries can expand (dilate) or constrict in response to changes in body conditions, including body temperature.

Human skin temperature at the surface is usually around 33°C (91°F). This temperature can be referred to as the skin temperature's "set point."

Q12. Would you expect skin with higher vascularity (containing more blood vessels) to return more quickly to the set point after a temperature change than regions with less vascularity? Why?

You are going to compare skin temperature recovery in the upper arm and the face.

Q13. Make a prediction about which body region (arm or face) will have the higher speed of temperature recovery.

The instructions for this experiment are given in the lab sheets provided by your instructor.

2

Instructions

To calculate rate of recovery:

1. Note the skin temperature after icing and when your team begins taking the readings.
2. Find the end temperature after 2 minutes of warming.
3. Subtract the starting temperature from the ending temperature (TF - TI) to get a change in temperature during measurement.
4. Divide the temperature change by 120 seconds to determine the rate of recovery in oC/sec.
5. Add your results to Table 2.2.

Table 2.2. Skin Recovery Data

Body region	Baseline Temp °C (before icing)	Starting Temp (TI) °C (after icing)	Ending Temp (TF), °C (2 min. after icing)	Rate of recovery (°C/s)
Upper Arm				
Cheek				

Q14. Did each area of skin return to baseline temperature in the 2 minutes that you took readings? If not, use the data to estimate how long temperature recovery would take.

6. Report your calculated rates of recovery to the data table on the board or the overhead.
7. Once you have the data for the other groups, calculate the class average in Table 2.3. You will graph the class averages on the next page.

Table 2.3. Class Averages

	Rate of recovery (°C/s)						
Body Region	Group 1	Group 2	Group 3	Group 4	Group 5	Group 6	Class Average
Upper arm							
Cheek							

To determine the average, add up the rates of recovery, then divide by the number of data points (groups).

For example: To determine the average of the data set: 5, 8, 9 & 6

- ❑ Sum the data = 5 + 8 + 9 + 6 = 28
- ❑ Divide the sum by the number of values = 28 / 4 = 7
- ❑ The average is 7

2

Q15. Which area of skin had the fastest rate of recovery? Compare the results to your original prediction. Based on this, which area has a higher density of skin capillaries?

Things to Note in Beginning Your Graph:
The data represent two different groups, not a series of continuous measurements. Discontinuous data calls for a bar graph.

8. The dependent variable is the variable that you measured or calculated from measurements. In this instance, the dependent variable is the rate of recovery of skin temperature. *The dependent variable goes on the y-axis.*
9. The independent variable is the variable that we predicted would cause a change in the dependent variable. In this experiment, the independent variable was the area of skin that was tested. *The independent variable for this graph will go on the x-axis.*
10. Give the graph a title.
11. Label the axes so the reader knows what they represent
12. Number the y-axis in as even divisions as you can.
13. On the graph in Figure 2.4, plot the class data for the average rate of recovery for the upper arm and the cheek.

Figure 2.4. Plotting class data for the average rate of recovery.

Graph Title:

Activity 2.3. Plant Homeostasis Using Stomata

As living things, plants must exchange gases, transport water, and maintain homeostasis, just like animals do. One structure plants use for this is called a stoma. A stoma is a regulated opening in the plant's leaves. When the stoma is open, gases and water can enter or leave the plant. When the stoma is closed, these cannot pass through the plant.

Procedure A. Examining a Stomata

This experiment will have you examine the opening and closing of *stomata* (*pl. of stoma*) in the plant *Tradescantia zebrina* "inch plant." You will examine the stomata's activity under a few environmental conditions and compare them to a control.

Instructions

1. Use scissors to cut two leaves from a plant as indicated by your instructor.
2. Place one leaf abaxial (purple) side up under the dissecting microscope.
3. Once the leaf is in focus, increase magnification to verify that the stomata are fully open (see Figure 2.5).

Figure 2.5. A diagram of the bottom of a leaf (abaxial side).

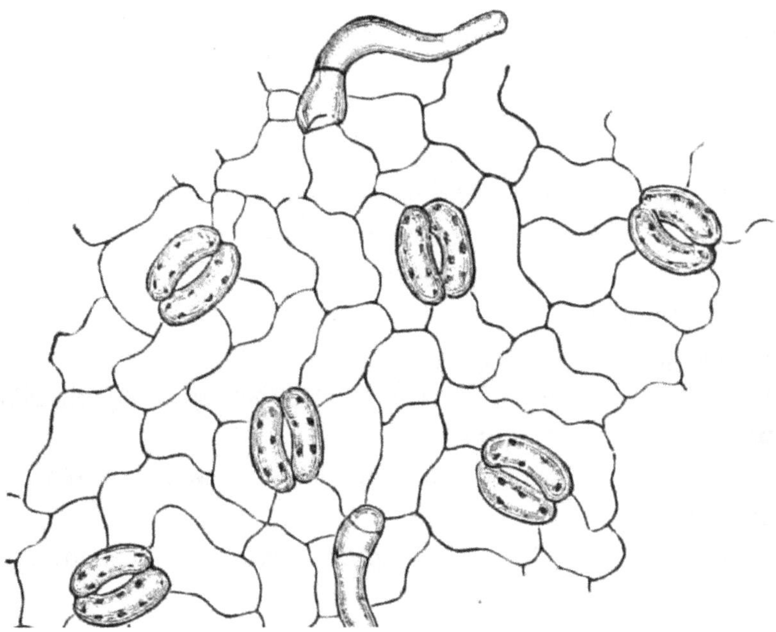

Use colored pencils to color code the diagram:

1. Color the stoma black
2. Color the guard cells green
3. Color the leaf dermal cells purple

Activity 2.4. Stomata Responses and Plant Pathology

Plant pathologists (scientists who study how plants respond to disease) discovered that stomata play a role in plant health and infection.

Q16. How might the ability of a pathogen (disease-causing organism) to infect a plant be affected by open stomata versus closed stomata?

Procedure A. Mimicking a Pathogen

For this part of the experiment, we will mimic a pathogen entering a plant using a chemical (H_2O_2) that plants secrete when being attacked.

Safety:

- ❑ Wear goggles at all times during this part of the lab.
- ❑ Use forceps (tweezers) to contact the liquids.

Instructions

1. Take the two leaves you cut from the plant.
2. Get two containers from your instructor marked A and B. They will look the same, but one represents exposure to the pathogen; one represents a control.
3. Place one leaf in each solution, purple (abaxial) side down.
4. Place the lid on the container.
5. Let the leaves soak for 15 minutes.
6. Use forceps to remove a leaf from the solution. Use forceps when removing the leaves to avoid temporary discoloration of fingers.
7. Place the leaf in a dry petri dish or weigh boat and view the abaxial side of the leaf under the microscope.
8. Sketch what you see in Figure 2.6. *Be sure to label the stomata, guard cells, and leaf cells.*

Q17. What differences do you see between the two leaves? Draw them in Figure 2.6.

Figure 2.6. Draw the differences between the two leaves.

Solution A | **Solution B**

Q18. Based on your observations and knowledge of stomata, which leaf represents the control, and which was "exposed" to the pathogen?

Q19. Explain using your understanding of homeostasis why the plant's stomata behaved as they did in the "exposed" treatment.

Q20. Are the stomata demonstrating positive- or negative-feedback homeostasis? Explain.

Name: ____________________ Lab Time: ____________ Due: ______________

2 Applying What You've Learned

1. Vasodilation of the skin allows the body to lose heat to the environment, while vasoconstriction reduces heat loss. What color would you expect the face of someone suffering from hypothermia (excessive loss of body heat) to be? Explain your answer.

2. A ripening apple releases a plant hormone called ethylene. Ethylene accelerates the ripening of unripe fruit in its vicinity so nearby apples also ripen. These, in turn, cause more ethylene release.

   - ❑ What do you predict will happen to the number of fruit ripening per unit time in this situation—increase, decrease, or stay constant?

   - ❑ What type of feedback mechanism does this example show?

   - ❑ What might be an evolutionary reason for this phenomenon of fruit ripening? That is, what advantage might it serve?

   - ❑ Sketch a graph in Figure 2.7 showing the effect of ethylene production on fruit ripening. (Make sure to label the axes.)

Figure 2.7. Graph showing the effect of ethylene production on fruit ripening.

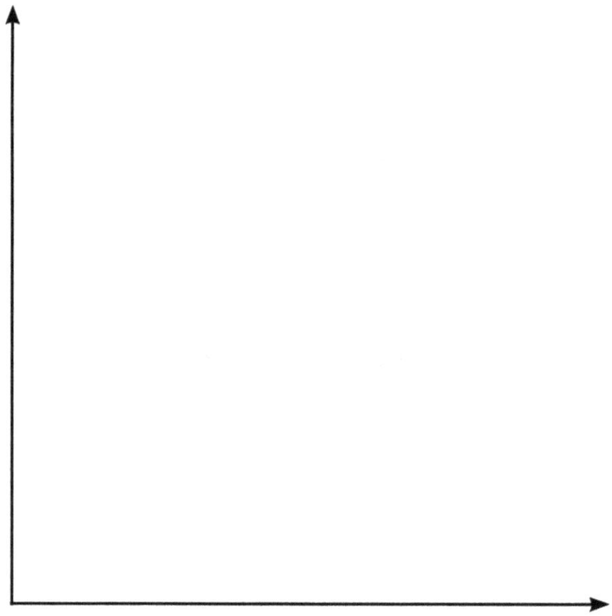

3. Diabetes is a (now) common disease caused by the body's inability to produce or use insulin. What would happen to blood sugar if a person had diabetes? Is this homeostasis? Why or why not?

Name: ____________ Lab Time: ____________ Due: ____________

3 Leaf & Synthesis Pre-Lab

1. In Activity 3.1, what are you using the syringe to do?

2. In Activity 3.1, which colors will the experiment test? Make a hypothesis about which of those colors will have the highest rate of photosynthesis.

3. What is a cuticle, and where is it located? See Activity 3.1.

3

4. What is the difference between glucose and starch? Which one is directly made during photosynthesis? Which one do plants store?

5. In Activity 3.2, what chemical will you add to the leaf to test for starch?

Name: ______________________ Lab Time: ____________ Due: ______________

3 Leaf & Synthesis

Photosynthesis is essential to all life, but only plants can carry out this process. **Photosynthesis** is the conversion of carbon dioxide into sugars using light energy. The equation for photosynthesis is shown below.

$$6CO_2 + 6H_2O \rightarrow \text{light} \rightarrow C_6H_{12}O_6 + 6O_2$$

Learning Outcomes

- ❑ Describe the basic chemistry of photosynthesis.
- ❑ Discuss factors that affect photosynthesis.
- ❑ Relate photosynthesis to plant and animal homeostasis.

Activity 3.1. Effect of Color (light quality) on Photosynthesis

(adapted from: Investigating photosynthesis with leaf discs from Science and Plants for Schools www.saps.org.uk)

The leaf is composed of layers of cells, as seen in Figure 3.1. The spongy mesophyll layer is normally infused with gases, oxygen, and carbon dioxide. Leaves (or disks cut from leaves) will normally float in water because of these gases. However, if you draw the gases out from the spaces, then the leaves will sink.

Figure 3.1. The cell layers of a leaf.

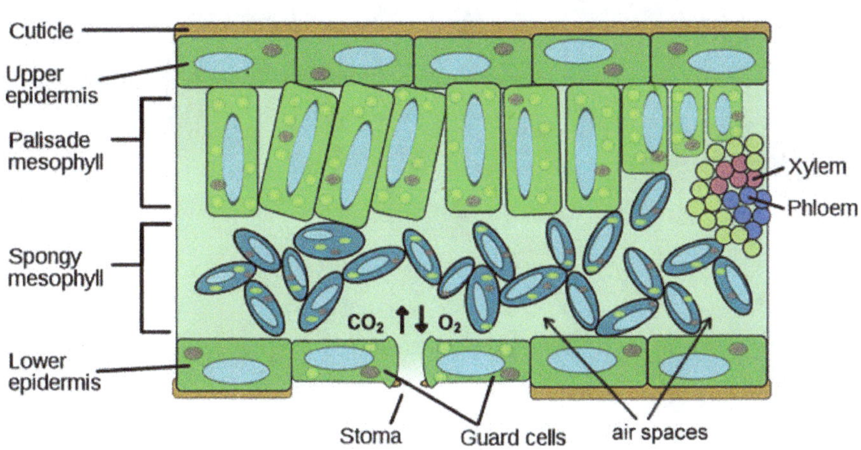

If this leaf disk is placed in a solution with an alternate source of carbon dioxide in the form of bicarbonate ions, then photosynthesis can occur in a sunken leaf disk. As photosynthesis proceeds, oxygen accumulates in the air spaces of the spongy mesophyll, and the leaf becomes buoyant and floats. The more photosynthesis there is, the more oxygen is produced. Oxygen and carbon dioxide are exchanged through openings in the leaf called stoma (or stomata).

Procedure A. Testing Colors

We will now test how the color of light (light quality) affects photosynthesis by monitoring oxygen production. You will test one color of light, then repeat the procedure for the other three colors.

See Figure 3.2 for a visualization of the following instructions.

Figure 3.2. A diagram illustrating steps in the color testing process.

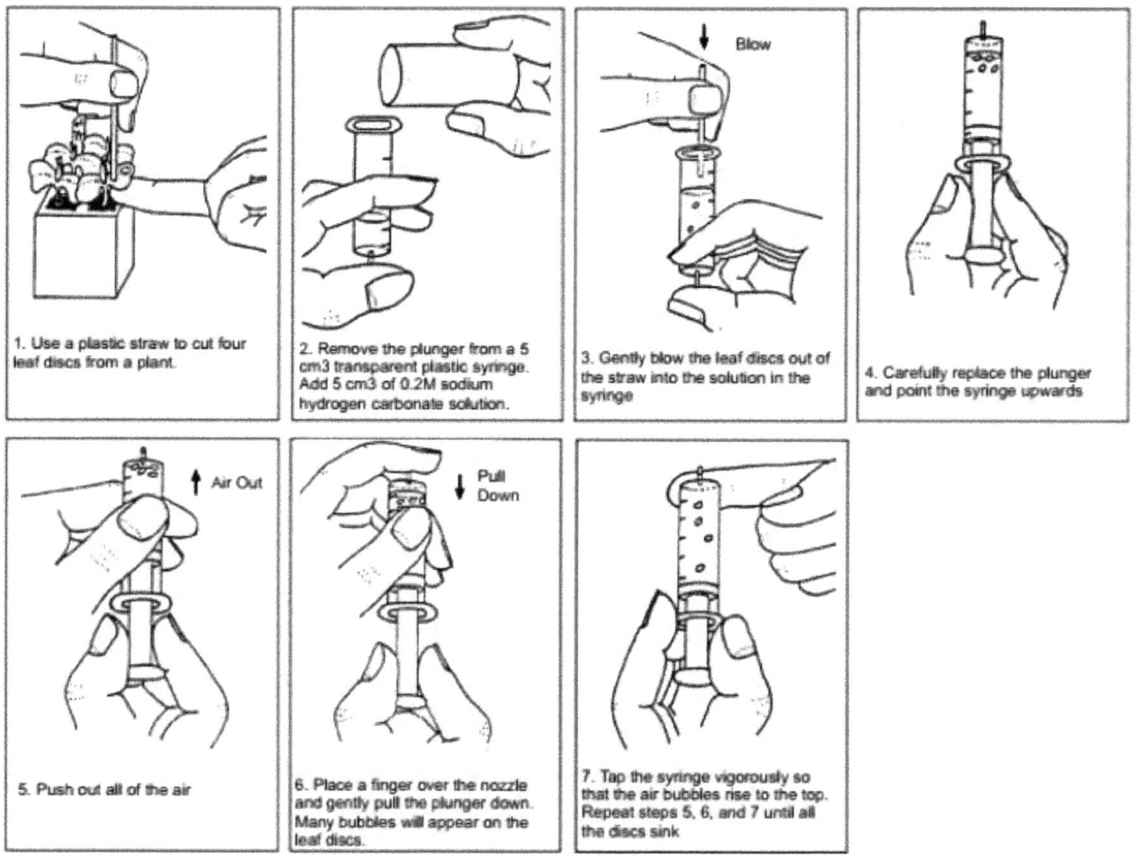

Instructions

1. Mix a small scoop (1/4 tsp) of sodium bicarbonate with 600 mL of water in your large beaker.
2. Add 1 drop of liquid dish soap to the bicarbonate solution and gently stir. If your solution generates excess suds, add more water and bicarbonate. This should be enough solution to run all four tests.
3. Hole-punch 10 leaf discs, ensuring they are uniform in texture and thickness and avoiding major leaf veins.
4. Place 10 discs into your syringe, then follow the instructions below to remove the trapped air. Be gentle when plunging the discs to prevent damage to the leaf tissue.
5. Pour the discs and solution into your small 150 mL beaker, then add enough bicarbonate solution to fill the beaker to 1/2 full.
6. Cover the beaker with the plastic film color that corresponds with the light color you wish to test. (Clear foil = white light, Red foil = red light, Green foil = green light, Blue foil = blue light)
7. Place the beaker directly under the light source (about 15cm away).
8. Watch the beaker and record the number of floating discs each minute for 15 minutes. Record your data in Table 3.1.
9. Upon completion, repeat steps 3-8 to test the other colors of light.

Table 3.1. Number of Leaf Discs Floating

time (min.)	discs floating in Clear (white) light	discs floating in Red light	discs floating in Green light	discs floating in Blue light
0				
1				
2				
3				
4				
5				
6				
7				
8				
9				
10				
11				
12				
13				
14				
15				

3

Q1. What color of light (clear, red, green, blue) had the highest rate of photosynthesis?

Q2. What color of light (clear, red, green, blue) had the lowest rate of photosynthesis?

Q3. What chemical produced during photosynthesis is making the leaves float?

Use the absorption spectrum of chlorophyll as illustrated in Figure 3.4, and your knowledge of the energy contained within different wavelengths of light to answer the following questions.

Figure 3.4. The absorption spectrum of chlorophyll.

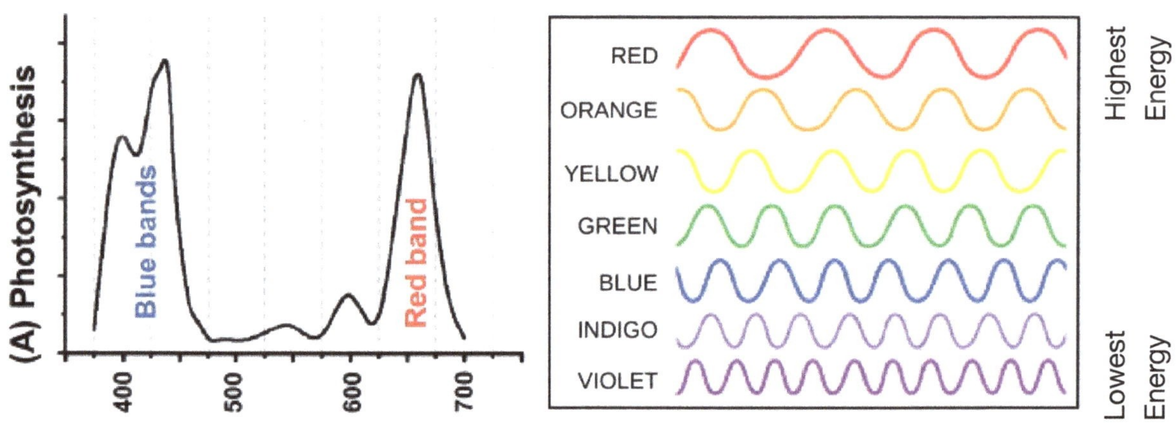

Q4. Based on the absorption spectrum of chlorophyll and the energy associated with each wavelength of light, which color(s) of light do we expect to see the highest rate of photosynthesis? What color(s) do we expect to see the lowest?

Q5. Using your bar graph, explain the results from your experiment. Which color(s) had the highest rate of photosynthesis (i.e., the most floating discs)? Which had the lowest?

Q6. Are there any results that do not agree with our predicted trends based on the absorption spectrum? Hypothesize why this might be so.

Q7. What would you expect to happen if you put a black cover (a film can) over the light or the beaker? Explain your answer.

Q8. The solution in the syringe contains sodium bicarbonate ($NaHCO_3$). Why does this solution need to contain a chemical with carbon in it? How does carbon relate to photosynthesis?

Q9. Given your answer above, what do you think would happen if you used distilled water for this experiment instead of the sodium bicarbonate? Why?

Activity 3.2. How Leaf Anatomy Reflects Homeostasis

(Adapted from: Nash, Young, & Inman. Growing a thicker skin: An Exercise for Measuring Organismal Adaptations to Terrestrial Habitats The American Biology Teacher (2015) 77 (6): 426–431. Micrographs provided by the author.)

You'll now integrate your understanding of tissues and their locations to the concept of homeostasis of the entire plant. It's easy to forget that all cells, tissues, and organs work together to maintain the whole organism, and are not just separate parts on a single slide.

Procedure A. Observing the Cross Sections of Leaves

For this activity, you will be observing cross sections of leaves. A cross section is a cut through a tissue that is perpendicular to the long axis. This type of cut allows you to see the inside tissues of a leaf as shown in Figure 3.5.

Figure 3.5. A cross-section cut of plant tissue.

"Outside World Above Leaf"

Section of Leaf = Inside Plant

"Outside World Below Leaf"

Most of the low-magnification photos show the entire leaf, with the outside world above and below. Higher magnifications, however, may show only part of the leaf and may cut off some of the structures seen in lower-magnification images.

The leaf consists of three types of tissues: dermal, ground, and vascular. Use Figure 3.1, the leaf diagram, to review the parts.

You will be looking for differences in these tissue layers to determine if the leaf belongs to a temperate plant, a desert plant, or an aquatic (water-dwelling) plant.

Procedure B. Examining Tissue Layers in Plants

Find the color photographs showing leaf slides from three different plant types. These are labeled as type X, type M, and type H. Each photo has a low-magnification view and a zoomed-in view.

Each set of photos has six different plants numbered 1-6. This makes 18 different choices.

Instructions

1. As a group, obtain one two-sided sheet of an **X-type**, an **M-type**, and an **H-type** plant.
2. Record the letter and number of each sheet in Table 3.4.
3. As a group, you'll need to make decisions about the following as you fill in Table 3.4:
    a. Which magnification should be measured or observed for a particular feature (low or high)?
    b. What units will you use? (millimeters or centimeters)

For the above decisions, think about how accuracy and precision reflect on science.

3

Table 3.4. Plant Leaf Observations

All measurements are made in the units of ____________________.

Plant	dermal thickness?	cuticle* thickness?	overall leaf thickness?	stomata** on top or bottom?	General observations?
X#					
X#					
M#					
M#					
H#					
H#					

A cuticle is a covering on the surface of the dermal tissue. It is a waxy substance. It is not a cell.

*****Stomata on these slides are a side view. You looked at them in your homeostasis lab as a surface view.***

Q10. Did thickness of the cuticle correlate to thickness of the dermal tissue? What about the overall leaf?

Q11. Is there a relationship between the cuticle thickness and stomata location?

3

Q12. Which plant type (X, M, or H) is most effective at preventing water loss through the leaf? What did you see that provides evidence for this?

Q13. Which plant type (X, M, or H) can function while floating on the surface of a pond? What did you see that provides evidence for this?

Q14. Based upon your conclusions above, which plant group represents:

- ❑ Plants that grow in hot dry conditions such as a desert.

- ❑ Plants that are aquatic.

- ❑ Plants that are in temperate (mild) conditions.

Before moving on, ask your instructor to identify the plants in your photomicrographs, then compare them to your hypotheses.

Q15. Were you correct or incorrect?

Q16. How does the anatomy of each plant type (X, M, & H) reflect homeostasis in that type's environment?

Q17. What were some sources of error in your data collection? How might an anatomist (a person who observes structures) reduce these types of errors if measuring data for a scientific paper?

Activity 3.3. Carbon Fixation

The goal of photosynthesis is to create sugars (carbohydrates) that plants can use to make energy and grow. The most common sugar produced from photosynthesis is called **glucose**. It is a six-carbon sugar formed from carbon dioxide using energy from the prior **light reactions**. Figure 3.6 shows the relationships between these substances.

Procedure A. Testing Boiling Substances on Leaves

Plants often store sugars as **starch**. Starch is a large molecule made of many sugars (glucoses) put together. It's called a **polymer** and is commonly an energy reserve of plants. Plants store starch in most of their cells, but can also store it in specialized structures such as tubers (e.g., potatoes).

Figure 3.6. An illustration of a leaf in different temperature environments.

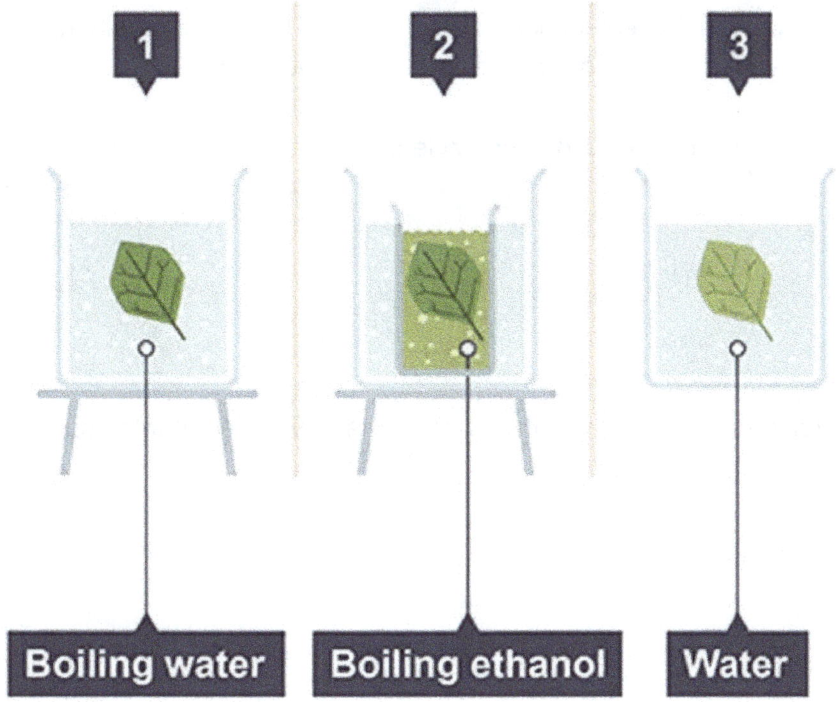

Instructions

1. Obtain one Coleus leaf and note where this leaf is green in color. Coleus is a variegated plant, meaning the leaves have different colors.
2. Take a picture of the leaf with your phone.
3. Fill a 250 mL beaker about half full with water.
4. Put the beaker on a hot plate and get the water to a boil
5. Submerge your leaf in the boiling water for 20-30 seconds. Do not overcook.
6. Carefully remove your leaf from the boiling water and turn off the heat.
7. Wear safety goggles for this next part.
8. Fill a large test tube with ethanol about halfway, just enough to submerge your leaf.
9. Place the test tube into the beaker on the hot plate as shown in the middle image above. The water should still be hot enough to affect the ethanol even if it's not boiling.
10. Heat the ethanol tube for 2 minutes.
11. Remove the test tube from the hot water.
12. Dump all the contents of your tube (the leaf and what should now be green liquid) into the hot-water beaker.
13. Using tweezers, gently submerge the leaf (which should now be clear) for about 10-15 seconds.
14. Carefully lift your leaf out of the hot water and place it flat in a Petri dish.
15. Add several drops of Lugol's iodine solution (IKI) onto the leaf until it's fully covered. It is important not to add too much excess fluid, as it will make the results hard to see. Leave for 1 minute.
16. With deionized water, gently rinse the leaf in the dish to remove excess iodine. It is ok to pour this solution down the sink or onto a paper towel.
17. Record what colors appear in the leaf–either tan (iodine color) or dark blue/black.
18. Compare the colors of your leaf now to the original picture you took.

Q18. What relationship do you see between the original leaf's colors and the iodine stain?

Q19. Lugol's iodine is specific for starch. Given this, what parts of your leaf contained starch?

Q20. What process made the starch?

Q21. What specific cell structure (organelle) and specific pigment is responsible for your answer above?

3

Q22. When the leaf was first boiled, the leaf stayed green. When it was in the ethanol, the leaf turned clear, and the ethanol became green. Based upon this, is the pigment soluble in water or ethanol?

Q23. Why would a plant produce pigments (such as orange or red) if they are not related to starch production? In other words, why might a plant produce pigments if not directly linked to sugars?

Q24. Plants grown in very low light produce a darker green pigment than those grown in higher light. Based upon your experiment, why might that be beneficial to a "shade plants" homeostasis?

3

Name: ____________________ Lab Time: ____________ Due: ______________

3 Applying What You've Learned

1. A plant nursery wants to maximize the rate of photosynthesis for their plants so they'll be tall when planting season arrives. According to the data collected during class, what colors and/or conditions should the plants be exposed to in order to meet this goal?

2. Some Internet sites say plants "make their energy from photosynthesis." In what way is that incorrect? In what way is it correct? Explain.

3. How would the dermal tissue of a cactus compare with that of a tropical houseplant? Give two differences.

4. Planting trees is said to "offset atmospheric carbon." Why would increasing the number of plants affect the amount of carbon in the atmosphere? Where would the carbon be going?

5. Based on your procedure from Activity 3.3, explain why you can cook vegetables in boiling water without them bleaching out.

3

Name: ______________________ Lab Time: ____________ Due: ______________

4 Plant Structure & Transport Pre-Lab

1. Examine the diagram of the compound microscope. What are the names of the two focus knobs?

2. Read the section "Getting started with a slide." What rule pertains to the coarse focus knob?

3. Where is a plant's dermal tissue system located? What about the vascular tissue system?

4. How long will you need to wait for Activity 4.3 to complete?

5. What parts of the plant will you sketch in Activity 4.4?

Name: ____________________ Lab Time: __________ Due: ____________

4 Plant Structure & Transport

4

As you learned in a prior lab, organisms are organized into systems, organs, tissues, and cells. While many people think these terms only apply to animals, they actually apply to all organisms, including plants.

In this lab, you will examine the anatomy of plant bodies and explore how the systems, organs, and tissues of a plant transport water from the roots to the leaves.

Learning Outcomes

- ❑ Safely operate a compound microscope.
- ❑ Identify the key organs and cells of plants.
- ❑ Discuss the role of transpiration in water transport in the xylem.
- ❑ Describe how changes in the environment alter the rate of transpiration.

Activity 4.1. The Compound Microscope

The parts of the microscope are shown in Figure 4.1.

Figure 4.1. Diagram of a compound microscope.

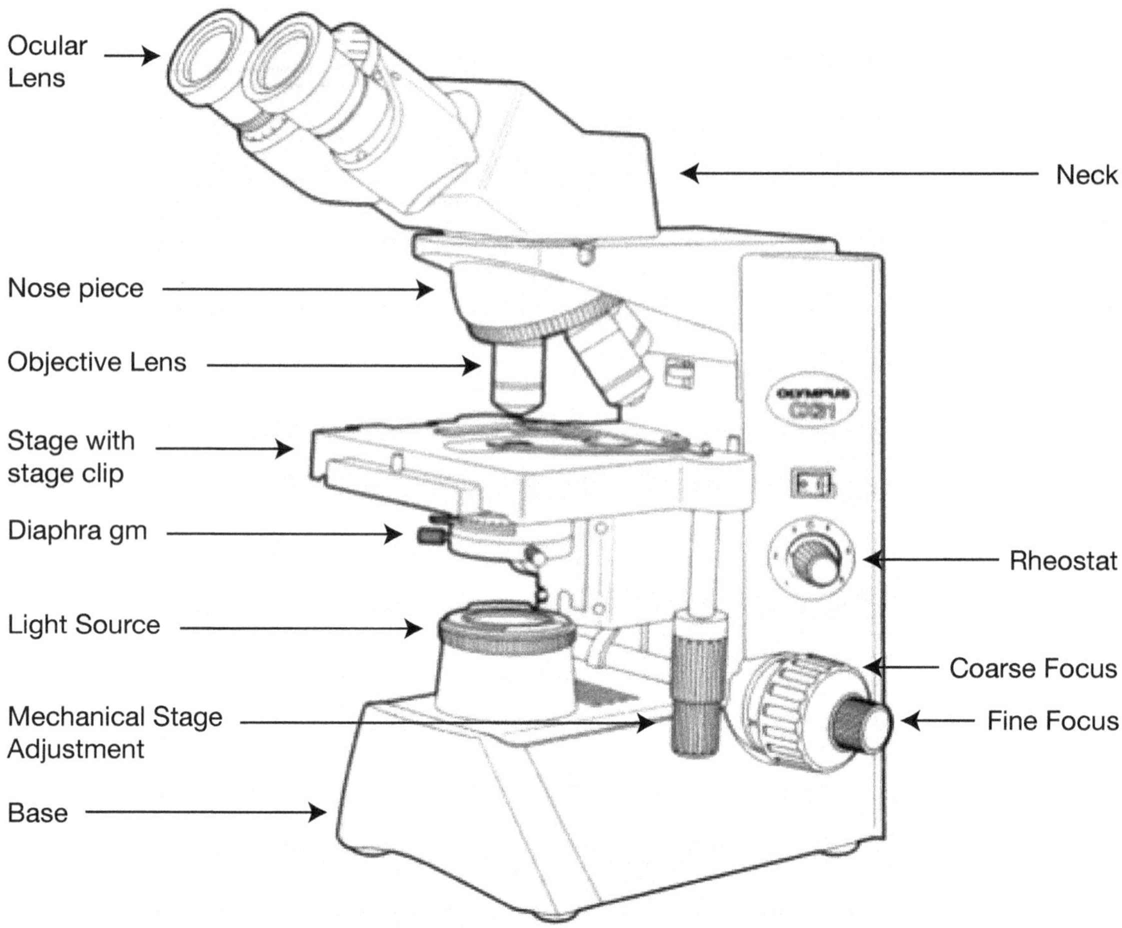

As your instructor discusses the microscope, describe the function of each part in Table 4.1.

Table 4.1. Microscope Parts and Functions

Microscope part	Function
Ocular Lens	
Objective Lens	
Stage w/ stage clip	
Mechanical Stage Adjustment	
Rheostat	
Iris diaphragm	
Coarse Focus	
Fine Focus	

Microscopes are complex instruments that require special care. The following procedures will help you comfortably use a microscope.

Sitting at Your Microscope

Microscopes are designed for people sitting with the microscope directly in front of them. Bending over to look in a microscope will hurt your back.

1. Place your microscope directly in front of you.
2. Adjust the height of your chair so that you can easily look into your microscope.

4

Cleaning Your Microscope

Although you should never touch the lenses (objectives or oculars) with any part of your body or with chemicals, sometimes it happens, and you need to clean your microscope.

1. Obtain a piece of lens tissue from the drawer located above your microscope.
2. Cover the tip of your pointer finger with the paper and use it to wipe the lens in a circular motion.
3. If the lens is not clean after this procedure, move to a new section of the lens paper, add a few drops of the blue lens cleaning solution (also in the drawer above your microscope), and wipe again in a circular motion.
4. If this procedure does not remove the debris from your lens, ask your instructor for assistance.

Getting Started with a Slide

1. Plug your microscope in and turn it on
2. Lower the stage completely. Click the low power objective into place.
3. Place the slide between the feet of the slide holder on the stage, then push it to the back of the holder.
4. Use the coarse adjust knob to focus the image you see through the oculars.
5. Use the fine-adjust knob to further sharpen your focus.
6. Adjust the iris diaphragm to find the right amount of contrast (this is usually pretty far to the right).
7. If necessary, increase magnification by clicking the next objective into place.
8. Use the fine-adjust knob to "fine tune" the focus. *Never use the coarse adjust knob at medium or high power*. Use the iris diaphragm to readjust the contrast as needed.
9. If necessary, increase magnification by clicking the next objective into place, sharpen your focus with the fine-focus knob, and adjust the contrast with the iris diaphragm.

Putting Away Your Microscope

1. Click the lowest-power objective lens (4X) into place and use the coarse focus adjustment knob to lower the stage.
2. Remove your slide and return it to the proper slide tray.

3. Turn off the light. Unplug the microscope and wrap the cord around the cord holder on the back of the microscope.
4. Return the microscope to its cabinet so that the back (not the oculars) faces you.

Hints for Looking at Specimens

1. If you lose your specimen, go back down to low power and start again. It won't take too long, and it will save you a great deal of frustration.
2. To avoid losing your specimen, make sure it is at the center of your field of view before you increase magnification.
3. If you can't find your specimen, go to the edge of the slide and locate the edge of the coverslip. Get the edge of the coverslip into focus. Your specimen should be at the same focal plane as the edge of the coverslip.
4. Do not use the coarse adjust knob at medium and high power. Not only do you risk damaging your slide and your microscope, but you are also likely to go so far out of focus that you will lose your specimen.

Procedure A. Calculating the Total Magnification

The total magnification can be calculated by using the following formula:

Total magnification = ocular lens magnification x objective lens magnification

Determine the individual and total magnifications of the following:

Q1. How much does the ocular lens magnify? ________ X

Q2. How much does each objective lens magnify? ________X ________X ________X

Q3. What is the total magnification at each objective? ________ X ________X ________X

Instructions

Letter E Slide

1. Find a Letter "E" slide in a student slide folder. These prepared slides have a printed letter "E" on them and are useful for learning to focus a microscope, since we all know what the letter "E" looks like.
2. In Figure 4.2, draw what you see in the field of view, and label it with the slide name and magnification. Remember that we always start with the lowest magnification lens and work our way up. Make sure to use the fine-adjustment focus only once you reach high power. These sketches can be pretty quick.

Figure 4.2. Sketch and label the Letter "E" slides.

Q4. Does the image appear right-side up, as you see it on the slide, or upside-down?

Q5. When you move the slide to the left, in which direction does the object appear to move within the field of view?

Q6. What about if you move the slide toward you?

Activity 4.2. Plant Tissue Systems

One unique characteristic of plants is the presence of **tissue systems**. These are cellular structures that extend throughout the plant body. Plants contain three different tissue systems: dermal, ground, and vascular. The **dermal tissue system** is represented by the plant's surface (*dermal* means "skin"). The **vascular tissue system** forms a series of continuous tubes usually located in the "core" of the plant. This tissue system functions to transport materials (*vascular* means "veins"). The space between the vascular tissue system and the dermal tissue system is filled by ground tissue. **Ground tissue** exhibits a variety of functions including storage, support of plant structure, and photosynthesis.

In this activity, we will look at the arrangement of these tissue systems in the plant stem and then apply what we have learned to the leaf and root.

Procedure A. Tissue Systems in Plant Stems

In this activity, you will locate and sketch three different cell types. Before you do that, you must first orient yourself to the organization of a dicot root.

Figure 4.3 is an example slide showing a cross-section of a dicot root.

Your slides may be slightly different, but the orientation will be the same.

Figure 4.3. Cross-section of a dicot root.

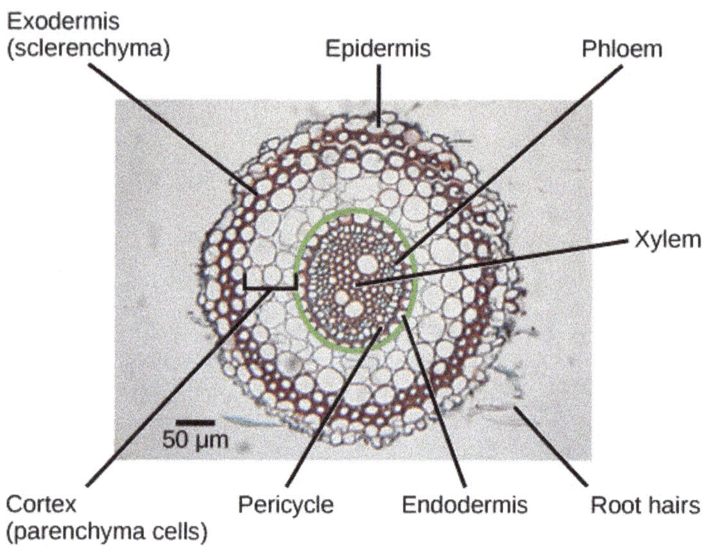

Instructions

Find the slides labeled "*Monocot*" and "*Dicot Root*" in your student slide folder. You can use the dicot root for this procedure.

1. Find the *outermost* layer of cells (i.e., "epidermis" on the slide image).

What type of tissue would these cells be? ______________________

2. Sketch one of these cells in the correct box in Figure 4.4.
3. Look in the *very center* of the root (i.e., "xylem & phloem" on the slide image)

What type of tissue would these cells be? ______________________

4. Sketch one of these cells in the correct box in Figure 4.4.

5. Look at the tissue between the two areas mentioned above (i.e., “cortex”).This part of the root is made up of ground tissue cells.
6. Sketch one of these cells in the correct box in Figure 4.4.

Figure 4.4. Sketch each type of tissue cell.

Dermal system cell	Vascular system cell	Ground system cell

Activity 4.3. Transpiration

In this experiment, you will observe the process of transpiration in celery stalks. **Transpiration** is the movement of water up the plant in the vascular tissue due to evaporation and certain properties of water (such as cohesion and adhesion). This process is illustrated in Figure 4.5.

Figure 4.5. An example of transpiration in a celery stalk.

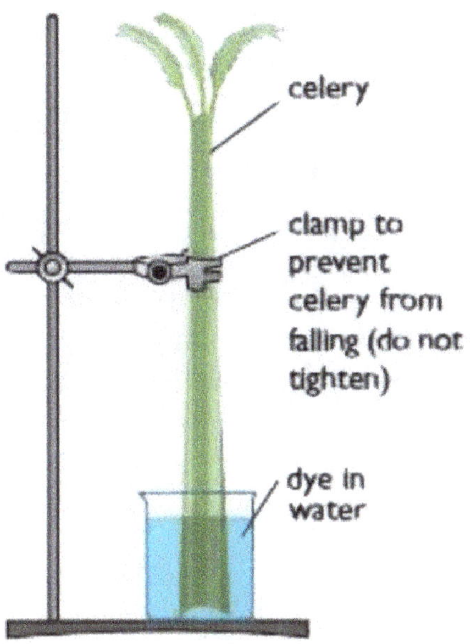

Instructions

You will do this activity as a *table group*.

1. Each table will need four stalks of celery and four 250 mL beakers. Try to pick similar celery for all four stalks.
2. Pour about 3 cm of water into the bottom of four 250 mL beakers
3. Add enough drops of red food coloring to make the water in each beaker bright red.
4. Cut the end off each celery using this procedure:
   a. Take the celery to the provided tubs which contain water.
   b. Hold the celery stalk underwater.
   c. Use a razor blade to slice about 2 cm from the top and 2 cm from the bottom of the stalk.
   d. Quickly transfer the stalk from underwater to the beakers.
5. Gently pat excess water from the top of the stalk with a towel or paper towel.
6. If you need to support your celery, use a ring stand as shown above.
7. Place each of your beakers of celery into one of the four following conditions:
   a. leave one beaker on the counter = "room conditions:
   b. put one in front of the fan = "wind conditions"
   c. put one in front of the bright light = "sunny conditions"
   d. put one under a damp plastic bag = "humid conditions" (use a rubber band to create a seal between the bag and the beaker)
8. Make a prediction about the rate/height of the dye in each celery.
   a. Which celery should have the highest transpiration rate?
   b. Which celery should have the lowest transpiration rate?
9. Leave the stalks in place for 30-45 minutes.
10. Over this time period, monitor your stalk ~ every 10 minutes or so. It may be necessary to moisten the cut end of the celery stalk or to scrape it with your knife. This prevents the ends from "healing," which can stop transpiration.
11. If the food coloring reaches the top of one of the stalks before the time is up, you may stop the experiment.

Be sure to stop the experiment for all four stalks at the same time.

12. After all are finished, remove each stalk from the beaker.
13. Carefully slice each stalk into 1-cm segments, starting at the top. As you make each slice, use a good light (your cellphone flashlight works well) and a magnifying glass to check to see if the dye appears in the veins. It will look something like Figure 4.6.

Figure 4.6. Cross-section view of a celery piece under a microscope, magnified 8x.

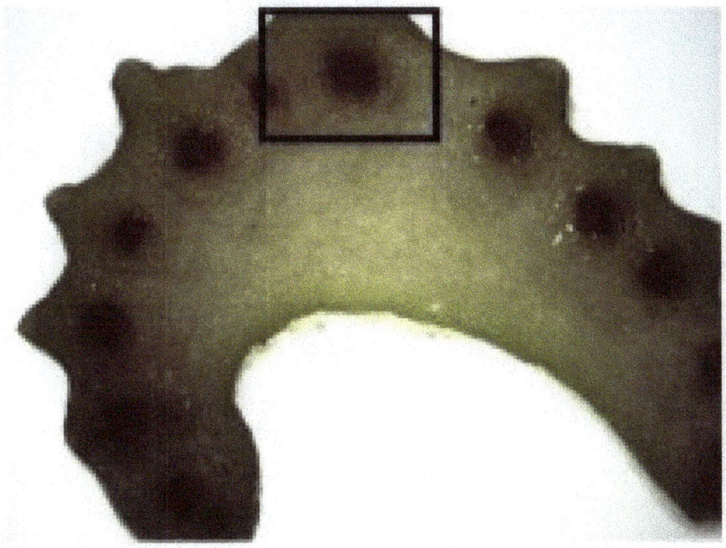

14. Continue slicing down the stem until you see ¾ or more of the stem's veins showing the dye.
15. Record the highest point the dye reached on each stalk, in centimeters, in Table 4.2.
16. Obtain measurements from the other lab tables, then average the results.

Table 4.2. Height of Dye in Celery (in cm)

	normal room	wind	sunny	humid
Table 1				
Table 2				
Table 3				
Average				

Did the average data support your predictions? Why or why not?

Q7. Relate the results of the experiment to what you have learned about vascular tissue and stomata in plants. How did these plant structures cause transpiration?

Q8. What part of the plant do you think has the greatest control over transpiration–the vascular tissue or the stomata? Why?

Q9. What were the sources of error in this experiment?

Activity 4.4. The Plant Body

The body of a plant is divided into two main systems called **shoots** and **roots**.

Q10. What do you think are the differences between these two systems?

Q11. What would be an organ of the shoot system? Do all plants have this organ?

Q12. Nodes are thicker, more noticeable areas on the stem. Nodes are where a bud was located. What is the function of nodes and buds in light of this system?

Procedure A. Shoot system

Instructions

1. Examine a young stem of asparagus or a similar plant provided.
2. Sketch the stem in Figure 4.7 and label the leaves, nodes, and internodes.

Figure 4.7. Sketch and label the stem.

Plant shoot

The first root that develops in a seedling is its **primary root**. A **secondary root** branches off of primary roots. Roots may have one main primary root (called a **tap root**) or have many small primary roots with many secondary roots also (called a **fibrous root**).

Q13. What would be an advantage to having a taproot system? What might be a disadvantage?

Q14. What would be an advantage to having a fibrous root system? What might be a disadvantage?

Q15. Based on the above, what type of root do you expect to see in plants that colonize sand dunes? Why?

Procedure B. Root System

Instructions

1. Get a seedling with washed roots from the plants available in the lab.
2. Sketch the root in Figure 4.8 and label the primary and secondary roots.

Figure 4.8. Sketch and label the root.

Plant root

Q16. What type of root (tap or fibrous) do you think this plant has?

Name: ______________________ Lab Time: ____________ Due: ______________

4 Applying What You've Learned

1. A photomicrograph of a xylem cell was taken at a total magnification of 400x. Based on this, what objective lens was used to take the photograph? Explain how you know.

2. Why do florists cut flowers with the stems underwater? You did the same thing with your celery. What would happen if they didn't do this? Relate your answers to cohesion.

3. Cells of the vascular tissue have stiff cell walls. Based upon what you learned today, explain why that is beneficial to transpiration.

4. What do you think happens to a houseplant's stomata if you forget to water it? How does this response help a plant maintain homeostasis until it is watered again?

5. Hydroponics is growing plants without soil, using only water solutions to carry minerals to the plants. Based on this, what conditions would a hydroponic grower want regarding light, heat, and humidity to ensure the plants receive enough water and minerals daily?

4

Name: ______________________ Lab Time: ____________ Due: ______________

5 Circulation Pre-Lab

1. What is the purpose of a circulatory system?

2. What is a "circuit" of a circulatory system? (see Activity 5.1)

3. Why can't you expose the worms to DI or tap water? (see Activity 5.2)

4. What chemical is the class going to test on the worms to determine changes in heart rate?

5. What makes ice fish's blood different from most other organisms' blood if exposed to Antarctic water? (see Activity 5.3)

Name: ______________________ Lab Time: ____________ Due: ______________

5 Circulation

The circulatory system of animals transports substances (nutrients, waste products, gases) throughout the body. Commonly, this system contains a pump (heart), vessels, and the fluid (blood) being pumped; however, this pattern is not required for all organisms.

In this lab, you'll compare circulatory systems in a few representative species and analyze the evolutionary and environmental factors evident in each.

Learning Outcomes

- ❑ Determine how blood and/or heart changes are affected by the environment.
- ❑ Discuss the evolutionary adaptations of species illustrated by their circulatory systems.

Activity 5.1. Vertebrate Circulation

All vertebrates have a closed circulatory system, meaning blood always stays in vessels. However, even though they all have the same type of circulatory system, there are differences between different groups, such as between birds and amphibians. Scan QR Code 5.1 or use the URL (go.chemeketa.edu/vertcirc) to view the animations.

QR Code 5.1. The Vertebrate Circulatorium

Instructions

Read the introductory material showing the key features of the animation. Then use the animation to fill in Table 5.1

Table 5.1. Fill in the Key Features of the Animation

	# hearts – one or many?	Ventricle type - single, partial, or separate?	Does blood mix – yes or no?
fish			
frog			
turtle			
bird			

5

Q1. The route blood takes leaving the heart and returning to it is called a circuit. Looking at the fish animation, how many circuits do fish have?

Q2. How many circuits does a bird have?

Q3. What might be an advantage and a disadvantage to having more than one circuit in an organism? Why?

Activity 5.2. Blackworm Circulation

(based upon: "As the worm turns" by Shannan Muskopf at BiologyCorner and Bohrer, K.E. 2006. in Tested Studies for Laboratory Teaching, Volume 27, Association for Biology Laboratory Education (ABLE).

This part of the lab will investigate how chemical or physical factors affect the circulation in a "black worm."

Background

Blackworms *(Lubriculus variegatus)* are freshwater worms that inhabit stagnant water, such as the edges of ponds, in North America and Europe. They feed on microorganisms and decaying vegetation in these habitats.

As seen in Figure 5.1, Blackworms have a **closed circulatory system**, in which blood is pumped by the tail muscles and transported in closed blood vessels. O_2 diffuses into the **dorsal blood vessel** across the worm's epithelium (surface) and then diffuses out of the blood into the worm's tissues. CO_2 is produced by the tissues, diffuses into the blood, and then diffuses from the blood to the air.

The heads and tails of blackworms are easy to distinguish. The head is more darkly pigmented and wider than the tail. In their natural habitat, blackworms can often be found with their heads buried in sediment and debris to help them find food. They keep their tails vertical, with the tail out of the water, to aid gas exchange.

Figure 5.1. The closed circulatory system of a blackworm.

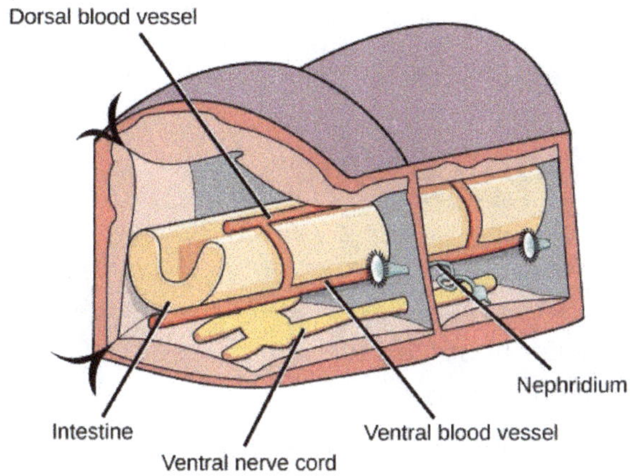

Q4. What direction do you expect blood to flow in the dorsal blood vessel–head-to-tail or tail-to-head?

You will be observing the **pulsation rate** of blackworms in the dorsal blood vessel. The dorsal blood vessel is large and easy to see because of the transparency of the worm's epidermis. The worms are long enough that several pulsation waves can be observed simultaneously along their length.

Because the worm's epidermis is so thin, some chemicals can diffuse through it into the blood vessels. This will allow you to expose the worms to different chemicals and observe their effects on pulsation rate.

Care and Handling of Worms

- ❑ Use only the provided water on your bench. *Never expose the worms to DI or tap water.*
  - » Tap water contains chlorine, which is toxic to the worms.
  - » DI is very hypotonic to the worms.
- ❑ Be careful when handling the worms. Sharp objects will damage them. Always use a pipette for transferring.
- ❑ The worms are sensitive to intense light exposure. Keep the microscope light low.
- ❑ Be patient in trying to view pulsations; it takes practice.

Scan QR Code 5.2 or use the URL (go.chemeketa.edu/blackworm) to watch a video of a blackworm pulsation.

QR Code 5.2. Pulse of Blackworm

Procedure A. Initial Observations of the Blackworm

Instructions

1. Fill each of your specimen bowls with worm water about 1/3 full.
2. Select three worms that are approximately the same size.
3. Use a plastic pipette to transfer your worm into your specimen bowl.
4. Examine your specimens and describe their appearance, how they are moving, or how they are grouped together. Identify the **anterior** (front) and **posterior** (tail) region of the worm.
5. Identify the **dorsal blood vessel**. Record your findings in the boxes of Figure 5.2.

Figure 5.2. Record and sketch your Blackworm observations.

Observations	Sketch one worm (with labels)

Procedure B. Baseline Heart Rate

You will be determining the pulsation rate of the dorsal blood vessel by recording observations from three worms. They will serve as your **control**. Observe one worm at a time.

Instructions

1. Using a pipette, gently transfer one worm into the trough of a well slide. Use a Kimwipe to remove excess water.
2. Place a coverslip loosely over the worm. Allow the worm to adjust to its new surroundings.
3. Place the slide on the microscope. Be sure to keep the light level low to avoid damaging the worms.
4. Identify a segment near the middle of the worm where the dorsal pulsation rate is evident.
5. Count the number of pulsations in that segment per minute. Record in Table 5.3.
6. Repeat for all three worms and then average the results.
7. When you have finished recording your observations, transfer your worm into a second container containing a small amount of the worm water with the disposable pipette.

Table 5.3. Baseline Worm Pulsations (beats per minute)

Worm 1	Worm 2	Worm 3	Average

Procedure C. Effect of Caffeine on Blackworm Pulsation Rate

Each group will be provided with a caffeine solution at a different concentration (0.1mM, 1mM, 5mM) to expose your worm to. Caffeine affects nerves, which should in turn affect muscles and, consequently, pulsations.

1. Before you test your worms, identify the following factors regarding your experimental design. Record those factors in Table 5.4.

Experimental design is the process of conducting research in an objective and controlled manner. It involves making a **hypothesis** (an educated guess) to predict the outcome of a study. Identifying the **independent variable** (the factor the researcher manipulates) and the **dependent variable** (the factor that is measured). As well as constructing a **prediction**. Predictions are generally written as "If…, then…, because…" statements. (If we do this… then we expect this… because we think this…)

Table 5.4. Identify the Experimental Factors

Hypothesis?	
Independent Variable?	
Dependent Variable?	
Prediction?	

5

Q5. How many worms should you test for your caffeine solution?

Q6. What are your controls?

Procedure D. Data Collection

Instructions

1. Perform your experiment and record your data in Table 5.5.
2. To expose your worm to caffeine, place approximately 2 drops on one side of your coverslip, then use a paper towel and capillary action to pull the solution under the coverslip.

Table 5.5. Caffeine Exposure Worm Pulsations (beats per minute) ________mM caffeine

Worm 1	Worm 2	Worm 3	Average

3. Share your average pulsation rate with other groups and fill in Table 5.6.

Table 5.6. Effect of Caffeine on Average Pulsation Rate

Groups	Control (no caffeine)	0.1 mM caffeine	1mM caffeine	5mM caffeine
1				
2				
3				
4				
5				
6				
Average				

4. Calculate the *change* in pulsation (caffeine pulsation - control pulsation) for each treatment. Record in Table 5.7.

Table 5.7. Change in Pulsation Rate Versus Control

	Average pulsation for caffeine (Table 5.4)	Average pulsation for control (Table 5.4)	Change in pulsation (+ / -) (beats per minute)
0.1 mM caffeine			
1mM caffeine			
5mM caffeine			

5. In Figure 5.3, plot the *change* in pulsation recorded in Table 5.5. Don't forget to label the axes and create a title.

Figure 5.3. Plot the changes in pulsation recorded in Table 5.5.

Graph Title:

Q7. What is your conclusion?

Q8. Did all the worms respond in the same way? Why or why not? What factors might influence individual response?

Q9. Were there any behavioral changes in the worms due to caffeine exposure? Did these behavioral changes correlate with the concentration of caffeine?

Q10. Would you classify caffeine as a depressant or a stimulant? Why?

Activity 5.3. Independent Investigation

Instructions

1. Repeat your experiment using a different chemical factor. This may include ethanol, acetaminophen, glucose, cinnamon, vanilla extract, and/or another choice provided by your instructor. You will only be working on *your team's worms*, so you do not need to compile group data from the rest of the class.
2. Discuss with your team how you will set up your experiment and collect data. Record in Table 5.8.

Table 5.8. Effect of ________________________________ on Worm Pulsations (beats per minute)

Worm 1	Worm 2	Worm 3	Average

Now calculate the change in pulsations from your control worms. Record in Table 5.9.

Table 5.9. Change in Pulsation Rate Versus Control (change in pulsation = factor tested - control)

Factor tested?	Average pulsation for factor tested (Table 5)	Average pulsation for the control (Table 3)	Change in pulsation (+ / -)

Q11. What is your conclusion?

Q12. How does your factor compare to caffeine? What might be a biological reason for this?

Activity 5.4. Blood

Mammals use blood to transport materials (nutrients, gases) throughout the body. Blood consists of cells (red blood cells) floating in a liquid (plasma). Red blood cells (RBCs) contain a protein called hemoglobin that binds to oxygen for transport, allowing RBCs to deliver oxygen to the body's tissues. The environment (e.g., temperature, water availability) can affect the condition of the blood. For this part of the lab, we'll investigate how blood composition differs under extreme conditions by focusing on a type of fish known as "ice fish" (Channichthyidae) found in Antarctic waters.

Procedure A. Environmental Effects on Blood.

(adapted from: https://www.biointeractive.org)

Most animals could not survive in the ocean around Antarctica. The average water temperature there is –1.8°C (28.8°F). At such low temperatures, you would expect most fish's blood to freeze. However, icefish (notothenioids) thrive in these cold waters. So, how do they do it?

This activity contains representations of the blood of a "normal" fish and blood of a notothenioid to compare.

Instructions

1. Fill two small beakers about ¾ full with crushed ice. Mix 1 teaspoon (8 grams) of rock salt into each beaker with a stirring rod. This will help the temperature in the cup get cold.
2. Using your wax pencil, label one test tube "A" and the other test tube "B."
3. Get samples of solutions A and B from your teacher. Fill one-third of test tube A with solution A and one-third of test tube B with solution B. Make sure that the test tubes are filled to the same level.
4. Put the test tubes into each cup of ice. Make sure that the liquids in both tubes are below the top of the ice. Let the solutions cool for 15 minutes.
5. While you are waiting, fill an additional beaker with ice and place your thermometer in the ice to cool. You do not need to mix rock salt in with this ice.
6. After 15 minutes, use the cooled thermometer to measure and record the temperature of solution A. Also note any visual changes, such as the formation of ice crystals in the solution.
7. Before measuring the temperature of solution B, rinse your thermometer with water and place it back on ice to ensure it is clean and cooled.
8. Record all data in Table 5.10.

Table 5.10. Temperature of Blood Tubes in Ice

	Temperature after 15 min.	Observations
Tube A		
Tube B		

Q13. Would the blood modeled by solution A or B be more advantageous for an animal living in a cold environment? Support your answer with evidence from your data or observations.

Procedure B. Ice Fish Biology and Evolution

To learn more about notothenioid blood, scan QR Code 5.3 or use the URL (go.chemeketa.edu/genes) to watch the BioInteractive short film The Making of the Fittest: The Birth and Death of Genes.

QR Code 5.3. The Making of the Fittest: The Birth and Death of Genes

As you watch the video, answer the following questions.

Q14. What is the "trade-off" between having red blood cells (RBCs) and being in freezing temperatures?

Q15. How do ice fish transport oxygen if they don't have RBCs?

Q16. How do antifreeze proteins keep fish from freezing?

Q17. Most fish, including the ancestors of notothenioids, do not produce antifreeze proteins. How did notothenioids "get" the gene for antifreeze proteins?

Q18. Explain the evolutionary process by which antifreeze proteins became common among notothenioids.

Q19. Explain the evolutionary process by which the hemoglobin gene is no longer in ice fish.

5

Q20. Explain how your two answers above tie into the video's statement that evolution "comes up with the best solution available" rather than the best possible solution.

Q21. Will species always evolve the adaptations they need to survive in their environment? Justify your answer.

Q22. Penguins also live in the Antarctic, but they do not have antifreeze proteins. What adaptations enable them to survive in a similarly cold environment?

Q23. Antifreeze proteins are an adaptation for extremely cold water. What would happen to an icefish's blood if it were instead placed in tropical (warm) waters? Why?

5

Name: ______________________ Lab Time: ____________ Due: ______________

5 Applying What You've Learned

5

1. Circle the correct term for each blank.
   Mammals have an [**open** / **closed**] circulatory system containing [**one** / **many**] heart(s) and [**one** / **two**] circuits.

2. Earthworms, frogs, and pigs all have hearts that pump blood throughout their bodies. However, they have some key structural differences. In a few sentences, describe how their hearts differ from each other.

3. If you were looking at an unknown blood vessel in a worm (for example), how could you tell if it was an artery or a vein? Give two possibilities.

4. You're doing an experiment about desert snakes. Based upon what you learned in this lab, what characteristics would you expect the snake's blood, heart, and vessels to have? Relate your answer to homeostasis.

Name: ____________________ Lab Time: ____________ Due: ______________

6 Gas Exchange & Metabolism Pre-Lab

1. What are two muscles that help with breathing in a mammal? Where is each located?

2. What does a pulse oximeter measure? What is the normal value for a person?

3. What is aerobic metabolism? Which gases are consumed and released in this process?

4. What do ectotherm and endotherm mean? Which word best describes an insect?

5. How long will you be measuring the cockroach in the jar?

Name: ______________________ Lab Time: ____________ Due: ______________

6 Gas Exchange & Metabolism

Gas exchange is a fundamental characteristic of all living things. To us, gas exchange is synonymous with breathing, but all organisms exchange gases even if they don't "breathe" in the way we think of the word. This lab will highlight how animals move gases into and out of their bodies (gas exchange) and what those gases are used for in the body (metabolism).

Learning Outcomes

- ❑ Understand how breathing affects gas exchange.
- ❑ Describe how respiratory rate changes with an organism's activity level.
- ❑ Relate gas exchange to an organism's metabolic rate.

6

Activity 6.1. Muscles & Breathing

Breathing (in mammals) is due to changes in the size of the chest cavity (thorax). The thorax changes in size due to muscle action. The two muscles most involved in breathing are the diaphragm and the intercostals. The **diaphragm** is a flat sheet located under the ribs and separates the chest from the abdomen. The **intercostal muscles** are located between the ribs and can move them closer or farther apart.

Procedure A. Exploring Your Own Muscles

In this procedure, you'll explore your own breathing and notice when these muscles contract and how that action affects the size of your chest cavity.

Instructions

1. Place one hand on your ribs and take a few deep breaths in and out. Notice what is happening to your ribs during inhalation and exhalation.

Q1. As you draw in a breath, how do your ribs move? Do you think the intercostal muscles are contracting or relaxing?

2. Place one hand on your abdomen and take a few deep breaths in and out. Notice what your abdomen is doing as you breathe.

Q2. What happens to your abdomen as you draw in a deep breath?

Q3. Do you think the diaphragm is contracting or relaxing?

3. Use QR Code 6.1 or the URL (https://tabletopwhale.com/2014/10/24/3-different-ways-to-breathe.html) to observe the respiratory structures of an insect, a mammal, and a bird. Then answer the next series of questions.

QR Code 6.1. Respiratory Structures

4. In what way are the airways of all three organisms structurally the same? List two.
5. In what way has a bird's airway evolved to be more efficient than a mammal's? How is this adaptation required for homeostasis in birds?
6. Do you think insects have a diaphragm? Why or why not? If not, how can they move gases in or out of their bodies?

Figure 6.1. Illustrates the difference between the respiratory systems of insects, humans, and birds.

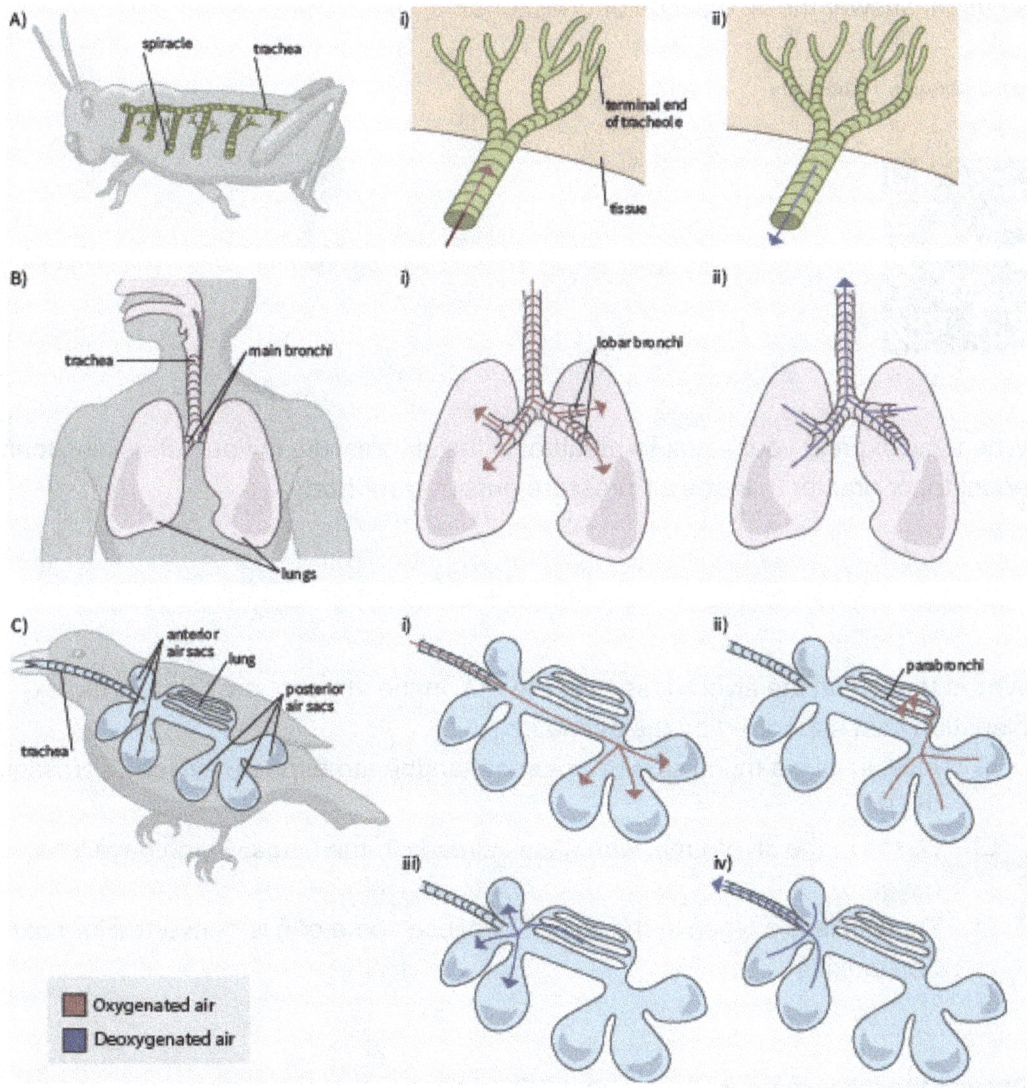

Activity 6.2. Oxygen in the Body

(Adapted from HHMI BioInteractive, Biology of Oxygen)

PO_2 - the partial pressure of oxygen is the pressure exerted by oxygen molecules. It is higher when there are more oxygen molecules in a given space. PO_2 varies at different points in the human circulatory system, enabling the movement of oxygen from areas of high concentration to areas of low concentration.

Instructions

Scan QR Code 6.2 or use the URL (https://media.hhmi.org/biointeractive/click/biology-oxygen/body-introduction.html) to watch the animation and learn how this works. As you watch the animation, answer the next series of questions.

QR Code 6.2. Oxygen in the Body

Q4. When you expand your lungs to breathe, is the air pressure in your lungs less than, equal to, or greater than the air pressure outside your body?

Q5. Why is the PO_2 in the alveoli less than the PO_2 in the atmosphere? Select the explanation best supported by the animation.

- ❑ Alveoli have a much smaller volume than the atmosphere, which decreases the PO_2.
- ❑ Gases in the alveoli mix with gases already in the lungs, which have less oxygen.
- ❑ There is less oxygen in the alveoli because some of it is converted into carbon dioxide.

Q6. Answer the following questions about PO_2 in the body.

- ❑ What is the PO_2 inside the alveoli? ____________________
- ❑ What is the PO_2 in the pulmonary capillaries? ____________________

Q7. Explain how the difference in PO_2 between the alveoli and pulmonary capillaries drives the movement of oxygen.

Q8. Explain how aerobic cellular respiration, the process by which cells use oxygen to get energy from food, affects PO_2 levels in a body's cells.

Activity 6.3. Blood Oxygen Saturation and Homeostasis

You can measure blood oxygen using a **pulse oximeter**. It is a device that clips onto a fingertip and shines a light through the finger. Sensors detect how much light is absorbed by tissues and blood, allowing the meter to calculate the percentage of oxygen in the blood.

Healthy people at sea level typically have oximeter readings between 95% and 100% oxygen saturation.

Procedure A. Using the Pulse Oximeter

Instructions

1. Choose a person from your lab group who is not wearing nail polish or artificial nails (as these can affect oximeter readings) as your subject. Have the person sit comfortably.
2. Turn the oximeter on and gently clamp it onto the tip of the subject's middle or ring finger (the skin on the pad of the index finger is more likely to be slightly thicker or callused from use).
3. Let the device remain on the finger for 30 seconds to calibrate and to allow the subject's rate to settle.
4. While you wait, measure the person's breathing. To do that, record how many breaths they take in 30 seconds, then multiply by 2.
   (An inhale and exhale together count as one breath.)
5. Record the readings below. Heart rate is also known as pulse.

Initial blood oxygen (%): ________ Initial heart rate (bpm): ___________

Initial breaths per minute (bpm): ________

6

Procedure B. Effects of Deep Breathing

Instructions

1. With the oximeter still in place, have the subject close their eyes, relax for 30 seconds, and then breathe slowly and deeply for 1 minute.
2. Count the number of breaths the subject takes during that minute.
3. At the end of one minute, record blood oxygen, heart rate, and breathing rate as before:

Blood oxygen after deep breathing: ________

Heart rate after deep breathing: ___________

Breathing rate after deep breathing: ________

Q9. What happens to blood oxygen, heart rate, and breathing rate after deep breathing? Why might this be?

Procedure C. Effects of Breath Holding

NOTE: If you get lightheaded easily, you may want another person in your group to be the subject.

Instructions

1. Record the original (resting) values from Procedure A in the first row of Table 6.1.
2. With the oximeter in place, have the subject take a deep breath and hold their breath for as long as comfortable. Notice what happens to blood oxygen during breath-holding.
3. After the breath is released, immediately record the data. This should be within about 5-10 seconds after release.
4. When the subject begins breathing again, record the heart and breathing rates every 15 seconds for 2 minutes.

5. Record the results in Table 6.1.

Table 6.1. Data from Breath Holds

Event	Blood oxygen (%)	Heart rate (bpm)	Breathing rate (bpm)
before holding breath (baseline)			
while holding breath (about 20 seconds)			
breathing resumes (within 5-10 seconds)			
15 seconds after resuming			
30 seconds after			
45 seconds after			
60 seconds after			
75 seconds after			
90 seconds after			
105 seconds after			
120 seconds after			

Describe how holding your breath affected the data. What might be a reason for this effect?

Q10. Did the subject's data return to normal within the two minutes that you measured? Which of these returned to normal the soonest?

Q11. Most likely, the changes in the data did not occur immediately upon holding or releasing your breath. Why not? How does this relate to the components of homeostasis from your first lab?

6

Q12. There are short-term and long-term sensors for blood oxygen levels. Which one do you think this experiment affected?

6

Activity 6.4. Gas Exchange and Metabolism

Gas exchange is critical to life because gases are used to create energy (ATP) in the body. In simple terms, the body makes energy by using oxygen to break down sugars. Carbon dioxide is created in this process. Often, this is called **aerobic metabolism** because gases are being used.

Because aerobic metabolism relies on oxygen and produces carbon dioxide, measuring the gases an organism uses or removes can estimate how much energy is being produced.

Procedure A. Measuring the Carbon Dioxide of a Roach

This part of the lab will focus on this relationship by measuring carbon dioxide production of giant cockroaches using CO_2 monitors and your phone.

Instructions

1. Use the instruction sheet provided by your instructor in the lab to download the Vernier software to your phone.
2. Only one of your group needs to use the software. The person with the software is responsible for collecting the data and sharing it with the rest of the team.
3. Weigh the empty flask (sensor jar) that will hold your cockroach. Record it as tare to the nearest tenth of a gram.

Tare (weight of flask) = ______________(g)

You will need to use this number later in the lab.

Procedure B. Baseline Data

Instructions

1. GENTLY put your cockroach into the sensor jar. Cockroaches are hardy but not invincible.
2. Weigh your cockroach & flask (together) and record the weight to the nearest tenth of a gram:

Weight of flask + cockroach = ______________ (g)

3. Determine the weight of the cockroach by subtracting the weight from step #2 from the tare on the prior page.

Weight of cockroach (alone) = ____________ (g)

4. Place the sensor into the mouth of the jar so it fits snugly.
5. Place your sensor jar on the lab bench and lay it on its side. This prevents a concentration gradient from forming in the jar. Use a stopper to support the sensor while the jar is on its side.
6. Make sure Vernier Graphical Analysis is actively reading CO_2 levels, and leave it undisturbed for 2 minutes to allow both the sensor and the cockroach to equilibrate.
7. After two minutes, select "Collect" to start collecting data. The first number collected will be the starting value.
8. Continue recording the CO_2 levels every minute for eight minutes. Use Table 6.2.
9. Calculate the change in CO_2 over this entire time.

change = final value - initial value

10. Use this value to calculate the rate of CO_2 production.

rate = change / total time)

11. Divide the rate (from Step 9) by your cockroach's weight to get a rate per unit of body mass.
12. Record in Table 6.2.
13. Take the sensor out of the jar and allow your cockroach to rest for ten minutes. Flush out the "stale air" out of the flask but avoid breathing into it.

Table 6.2. Baseline CO_2 of Cockroach

Temperature = __________ °C

Time	start	1 min	2 min	3 min	4 min	5 min	6 min	7 min	8 min	Change (ppm)	Rate (ppm/min)
CO_2											

rate / weight = ____________________ ppm/min*g

Activity 6.5. Effect of Cold and Heat

Repeat the experiment from *Activity 1, Procedure A*, using the same cockroach. This time, however, you will be changing the temperature of the cockroach.

Procedure A. Experimenting with the Temperature of the Cockroach

Instructions

1. Your instructor will assign you a temperature—either warm or cold.
2. Allow your cockroach, chamber, and CO_2 sensor to equilibrate to the new temperature for 5 minutes. During this time, the CO_2 sensor should be inserted into the jar, and the Vernier Graphical Analysis program should be running.
3. Measure the temperature of the new environment when you're ready to start data collection.
4. Just as before, record the initial CO_2 value and select "Collect" to start recording data.
5. Continue recording the CO_2 levels every minute for eight minutes. Make sure to record your data in the correct table below.
6. Calculate the change and rate as described previously. Divide by weight to determine the rate of CO_2 production per minute per gram of cockroach.
7. Take your jar out of the temperature environment and remove the sensor. Allow your cockroach to rest for 20 minutes. Extra time is needed for the cockroach to return to room temperature. Flush out the "stale air" out of the flask, but avoid breathing into it.
8. Record your data in Table 6.3 (cold) or Table 6.4 (warm), depending on what temperature condition you were assigned.
9. Once you have collected your data for one temperature condition, see if your instructor wants you to repeat another temperature.

Table 6.3. CO_2 of Cold Cockroach

Temperature = __________ °C

Time	start	1 min	2 min	3 min	4 min	5 min	6 min	7 min	8 min	Change (ppm)	Rate (ppm/min)
CO_2											

rate / weight = ____________________ ppm/min*g

Table 6.4. CO_2 of Warm Cockroach

Temperature = __________ °C

Time	start	1 min	2 min	3 min	4 min	5 min	6 min	7 min	8 min	Change (ppm)	Rate (ppm/min)
CO_2											

rate / weight = ____________________ ppm/min*g

10. Add your group's data to the data table at the front of the room.
11. Once all groups have contributed, copy the class data into Table 6.5. It's easier to interpret if you list the data by cockroach size (mass).

Table 6.5. Class Data: The Rate of CO_2 Consumption per Weight of Cockroach

		Rate (CO_2/min) per weight (g)		
Groups	Mass (g)	Room temp	Cold temp	Warm temp
1				
2				
3				
4				
5				
6				
Average				

Q13. Looking at the rate versus the mass, is there any trend in the rate as the cockroach gets larger or smaller? If so, what is it?

Q14. What effect did the warmer or colder temperature have on the CO_2 produced by the cockroaches?

Q15. Based upon your data, what temperature(s) required the most energy production (metabolism) by the cockroach? Why might that be?

Q16. Insects are ectotherms, meaning they require heat from the environment. Mammals are endotherms, meaning they produce heat internally. How would your data change if you were to put a mammal in the jar instead of an insect? Why might that be?

6

Name: ____________________ Lab Time: ____________ Due: ______________

6 Applying What You've Learned

1. What would happen to a person's lung capacity and possibly their blood oxygen if their chest was wrapped tightly with a bandage? Why?

2. A parasite called the tracheal mite (*Acarapis woodi*) can invade and lodge itself into the spiracles and tracheae of honeybees.

3. What would happen to a honey bee's foraging range (flight distance) if it were infected with these mites? Why?

4. How might long-term exposure to these mites affect a honey bee colony over time? Why?

5. If climate change causes temperatures to increase, what change in cockroach activity might you expect? Why? Use what you've learned in this lab to complete the negative feedback loop in Figure 6.2.

Figure 6.2. Components of a negative feedback loop.

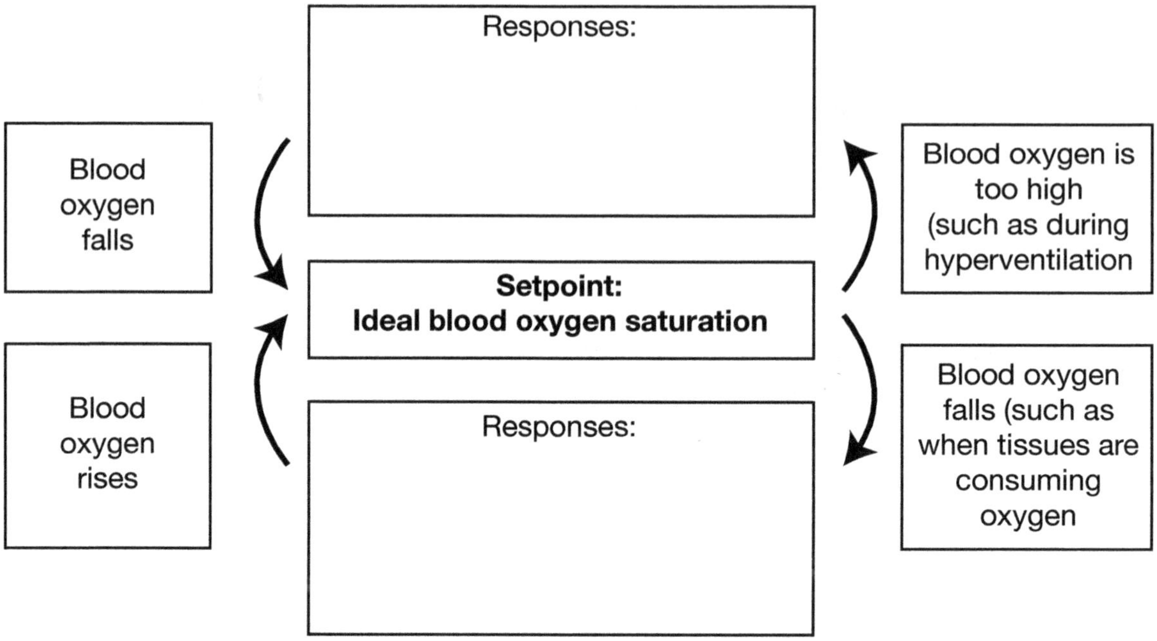

Name: ______________________ Lab Time: ____________ Due: ______________

7 Response to Environment Pre-Lab

1. What is the name for the cell used by the nervous system to send messages?

2. What is one difference between a reflex and a reaction?

3. Parts of your lab refer to the central nervous system. What organs are included in this?

4. How does a sensory neuron differ from a motor neuron?

5. What is a tropism? Which tropisms will you examine as part of today's lab?

Name: ______________________ Lab Time: ____________ Due: ______________

7 Response to Environment

A common characteristic of most animals is a nervous system. The nervous system uses specialized cells that communicate using electrical signals. These signals are used to integrate and coordinate the body, contributing to homeostasis.

Learning Outcomes

- ❑ Identify and describe the structures of a neuron, a nerve, and a spinal cord.
- ❑ Describe common reflexes and reactions.
- ❑ Describe common sensations and sensory perceptions.
- ❑ Explain the role of plant hormones in plant tropisms and how these relate to homeostasis.

7

Activity 7.1. Structure of the Nervous System

The functional unit of the nervous system is a **neuron**. Neurons are highly specialized cells that receive, interpret, and respond to a variety of stimuli. Neurons send nerve signals (impulses) using electrochemical changes in their membranes.

In a typical neuron, long branches (**dendrites**) lead to a large **cell body**. The dendrites receive information. The cell body integrates the signals it receives from the dendrites and produces a signal that is transmitted down the **axon**. At the end of the axon, the signal is transmitted to another cell, such as a muscle cell or a gland cell.

Instructions

1. Find the slide labeled "Motor Nerve Cell."
2. Use the microscope to focus on the neurons. Neurons are the larger, spidery cells.
3. Sketch a neuron in Figure 7.1.
4. Label: dendrite, cell body, axon.
5. Use arrows to indicate the direction of signals through the neuron.

Figure 7.1. Sketch and label the "Motor Nerve Cell."

Neuron
Total Magnification:

Q1. Why do you think the dendrites of this neuron are branched?

Q2. Why is it important that neurons have one-way flow of information?

A **nerve** is a bundle of neurons wrapped in connective tissue that enables communication between the brain and spinal cord and the body's periphery. The neurons in a nerve may contain a substance called **myelin**. Myelin is made of phospholipids (a type of fat) and acts as insulation for the neuron's axon.

6. Find the slide labeled "Teased Nerve."
7. In this case, you will see several long, skinny processes that appear segmented. The thicker regions represent myelin. The thinner regions are called nodes and represent areas where myelin is absent.
8. Sketch an axon covered with myelin in Figure 7.2.
9. Label: the axon, myelin, node.

Figure 7.2. Sketch and label the "Teased Nerve."

Teased Nerve
Total Magnification:

Q3. Neurons use ions (charged atoms) to create an impulse. Myelin is made of fat. Given this, why do the ions only create an impulse at the nodes?

Activity 7.2. Reflexes

A **reflex** is a rapid, involuntary response to a stimulus. For instance, if you put your hand on a very hot surface, you will instantly pull your hand away. Specialized neural pathways called reflex arcs control these reactions. Reflex arcs are primarily "local" in effect. That means that during a reflex, only a specific part of your body reacts immediately. Oftentimes, your brain may not be consciously aware that the reflex occurred.

A reflex arc starts with sensory **receptors** (sensors) that detect a specific stimulus. For example, some receptors detect heat while others detect pressure, touch, or pain. A nerve impulse travels from sensory receptors through a **sensory neuron** to the spinal cord. In the spinal cord, the impulse from the sensory neuron is routed to a **motor neuron**, which carries it to an **effector**, such as a muscle or gland. The effector then causes a response (usually, a movement). In some reflexes, another neuron (known as an **interneuron**) connects the sensory and motor neurons.

Procedure A. Diagram of a Typical Reflex Arc

Instructions

1. Observe the figure of a typical reflex arc (Figure 7.3) and complete the diagram by labeling the appropriate terms and pathways listed below it.

Figure 7.3. Typical reflex arc.

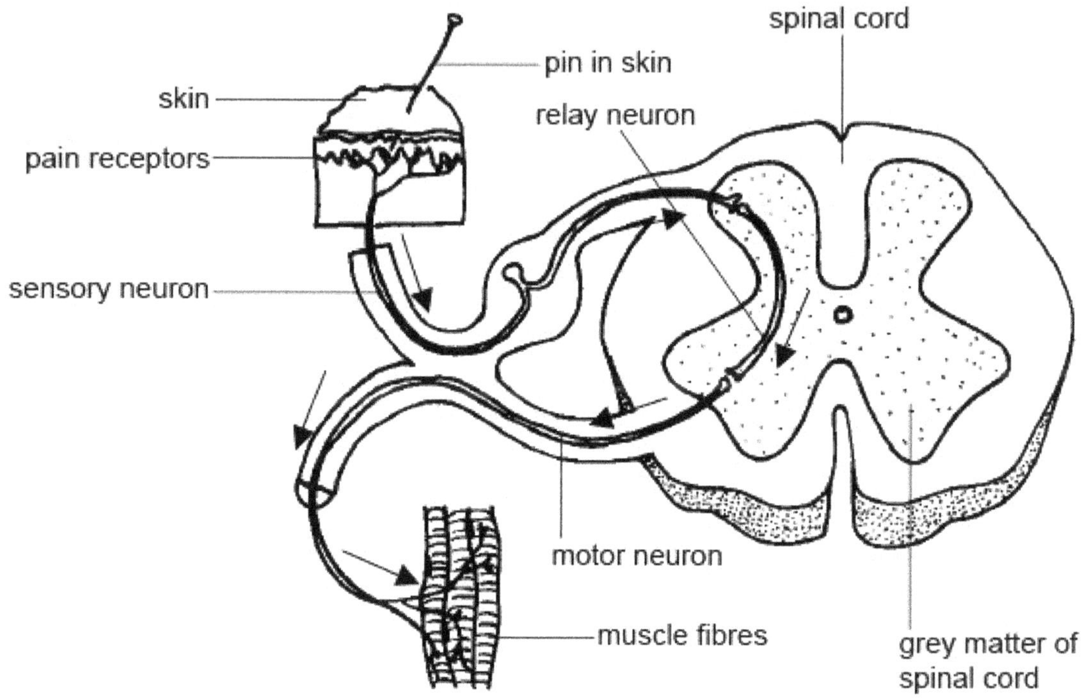

2. Label the neurons: sensory neuron, motor neuron, interneuron.
3. Circle and label: sensory receptor, effector.
4. Draw: arrows showing the direction a signal will travel.

Q4. What do you think the response will be in this diagram?

Activity 7.3. Patellar Reflex

The **Patellar Reflex** is a relatively easy reflex to observe. It's performed by striking just below the knee and observing the thigh muscles. The parts of this reflex are shown in Figure 7.4.

Figure 7.4. Components of the Patellar Reflex.

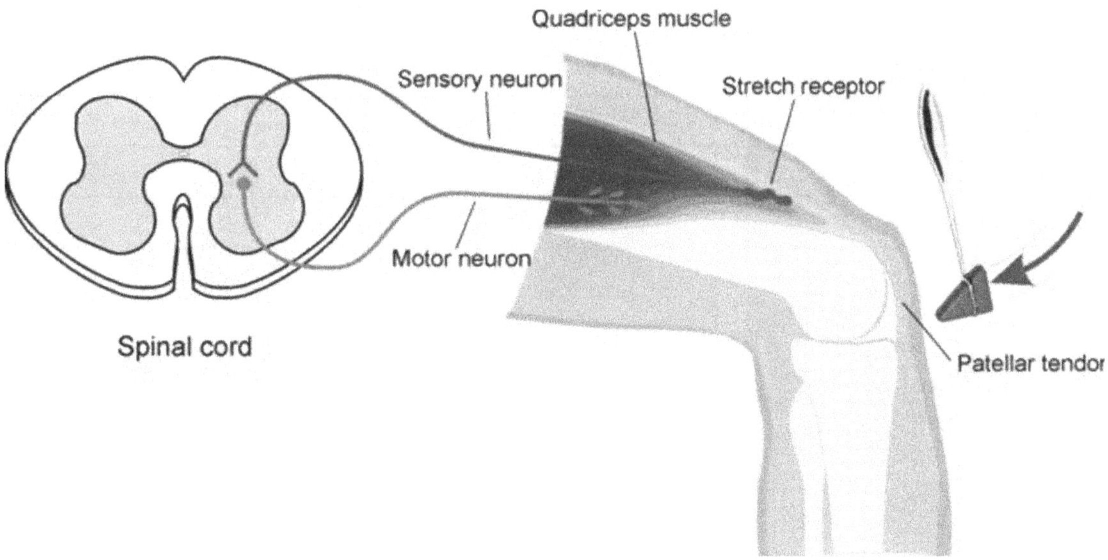

Q5. What part of the body will detect the hammer strike? What role of the reflex is this?

Q6. What part of the body will visibly respond to the hammer strike? What role of a reflex is this?

Procedure A. Testing Reflexes

Instructions

1. Work in pairs and each person collects data on themself and records it in the table below.
2. Sit on a table or desk high enough for your legs to hang freely.
3. Close your eyes.
4. Have your lab partner tap the tendon just below your kneecap (patella).
5. If no reaction occurs, try moving the point of impact around the knee.
6. Record your observations in Table 7.1 for each leg.
7. Repeat the procedure again, but with the changes indicated in Table 7.1.

Table 7.1. Results of Reflex Testing

Procedure	Right leg response?	Left leg response?
a. eyes closed		
b. eyes open		
c. keeping thighs clenched		
d. counting out loud		

Q7. Were you aware of your leg moving when your eyes were closed?

7

Q8. Did any of the changes in the procedure (b-d) affect either leg's response? Why or why not?

Q9. What might be an evolutionary reason to have this particular reflex? In other words, what's the point of contracting the thigh when the patellar tendon is stretched?

Activity 7.4. Learned Response

A response differs from a reflex by being (in general) voluntary (or learned). Responses and reflexes share similar components and physiology but usually differ in that responses are mediated primarily by the brain and are thus more complex.

Procedure A. Reaction Time

Instructions

1. Work in pairs and each person collects data on themself and records it in Table 7.2.
2. One group member should hold the reaction time stick at the top and let it hang vertically.
3. The test subject should position their thumb and forefinger at the thumb line at the bottom.
4. The subject can keep their eyes on the ruler but cannot touch it yet.
5. Without warning, drop the stick and have the subject try to catch it between their fingers as it falls.
6. Record the distance the stick fell in milliseconds (ms) by measuring the distance between the thumb line and the point where the stick was caught.
7. Repeat three more times and then average the results.

Table 7.2. Reaction Distance (in milliseconds)

	Trial #1	Trial #2	Trial #3	Trial #4	Average
Distance (ms)					

7

Q10. Did your performance "get better" over the four trials? Why might that be?

Q11. How does your answer to the question above illustrate a difference between a reflex and a reaction?

Procedure B. Hypothesis Testing for Reaction Times (using your non-dominant hand)

Could reaction times be linked to something physical about the subject? To test this, you'll perform an experiment.

Instructions

1. Combine with the other lab group at your table so you are now a group of 8.
2. Talk amongst your group about a factor you could investigate such as weight, age, sex, musicians, athletes, etc. Pick anywhere from 4-8 people to be the subjects for this experiment. The individuals selected must differ enough in the "factor to be measured" to matter.

3. Once you have decided on a factor to test, get it cleared by your instructor.

Our factor is: ______________________________

Q12. What is your hypothesis about the effect of this factor on reaction rate?

4. Repeat as before, except have the people use their non-dominant hand to perform the experiment. Record in Table 7.3.

Table 7.3. Factor Affecting Reaction Rate (in milliseconds (ms)

Factor of interest	Name	Trial #1	Trial #2	Trial #3	Trial #4	Average

5. Calculate the average reaction rate for each factor of interest using the data in 7.3. If you were testing two different factors, then you should have two averages you calculate here.
6. Record your calculations in Table 7.4.

Table 7.4. Record the Averages

	Factor of interest	Average reaction time (ms)
Group 1		
Group 2		

7. Make a graph of your average data in Figure 7.5.

Figure 7.5. Graph the average reaction times.

Graph Title:

Q13. Did the data support your hypothesis–yes or no? ____________

Q14. Interpret the data. What is a possible biological reason for your results?

Q15. What might be a source of error for your experiment? What could you do to improve your experiment?

7

Activity 7.5. Sensation

In the previous parts of this lab, you learned how the body responds to a stimulus via the nervous system. This part of the lab, however, will focus on how the nervous system detects a stimulus. This is called **sensation**. You will use a two-point test to determine how the skin detects touch, and to what level.

Procedure A. Recording Sensations

The areas tested will be: wrist, palm, back of the hand, cheek, back of the neck, and upper arm.

Instructions

1. Make a hypothesis about which area should have the smallest (closest) distance.
2. Work in pairs, and each person collects data on themself and records it in Table 7.5. Obtain a compass (caliper).
3. If using a digital caliper, zero it while closed. Set the readings to mm.
4. Start with the compass points spread out by about 25-30 mm (about an inch).

Caution: Compasses can be sharp. You do not need to apply very much pressure.

5. Have your subject close their eyes or look away so that they cannot see the compass touch the skin.
6. Touch the skin on the wrist with both points of the compass simultaneously.
7. Have the subject say if they feel both points clearly.
8. Gradually bring the points closer together and repeat the test until the subject stops feeling both points clearly separate and instead feels one point of contact.
9. If the subject only feels one point of contact at the initial setting, rather than moving the calipers closer together, move the calipers further apart until you can distinguish two separate points.
10. Repeat this procedure for each area in the data table.
11. Calculate the reciprocal value for each measurement. You'll use the data in Table 7.5 to graph your results.

Table 7.5. Two-Point Discrimination

	inside of wrist	palm	back of hand	cheek	back of neck	upper arm
distance in mm (smallest feeling both)						
reciprocal value = 1 / distance						

12. In Figure 7.6, graph your data as follows:
    a. The body part tested is the independent variable and goes on the X-axis
    b. The reciprocal value is the dependent variable and goes on the Y-axis (we use the reciprocal to keep all values between 0 and 1; it's easier to graph).
    c. Give this graph an appropriate title and label the axes.

Figure 7.6. Graph the two-point discrimination data from Table 7.5.

Graph Title:

Q16. Based upon your graph, what part of the body has the highest density of receptors? How did you know?

Q17. Mammals (such as a human) are bilaterally symmetric, meaning their body plan has a front (with a head) and a back. Does your data from the experiment show a relationship to this body plan? Explain.

Q18. How does your data from these experiments illustrate the nervous system's role in homeostasis?

7

Activity 7.6. Plant Tropisms

(adapted from a series of activities from the UK site Science & Plants for Schools https://www.saps.org.uk/)

While plants lack a central nervous system like animals, they do use electrical and chemical signals to respond to the environment. A **tropism** is the directional growth of a plant in response to external factors such as water, gravity, wind, or light. Usually, these tropisms are under the direction of plant hormones; however, we won't be looking at the hormones directly in this lab, only their effect.

Tropisms can be sorted into positive or negative.

- ❑ **Positive** is when the growth is towards the stimulus.
- ❑ **Negative** is when the growth is away from the stimulus.

Procedure A. Gravitropism

Instructions

1. Examine the plants in the gravitropism (gravity) demonstration area.

Q19. What type of gravitropism (positive or negative) is the stem showing?

Q20. What type of gravitropism (positive or negative) are the roots showing?

Procedure B. Phototropism

Instructions

1. Examine the plants in the phototropism (light) demonstration area.

Q21. Are the stems showing positive or negative phototropism?

Q22. Do you expect the roots to show the same type of phototropic response? Why or why not?

Q23. Why is phototropism essential for plant homeostasis?

Name: ____________________ Lab Time: __________ Due: ____________

7 Applying What You've Learned

1. Consider a nerve impulse traveling down a neuron. Do you think these cells use a lot of ATP or very little? Explain your answer.

2. Compare the patellar reflex with the ruler drop reaction. Where in the central nervous system are the two responses processed?

3. Figure 7.7 is a graph showing reaction time versus the intensity (or strength) of a stimulus, similar to what you did in the two-point test. What does this graph show, and how could you use this information in your day-to-day life?

Figure 7.7. Reaction time vs. stimulus intensity (strength).

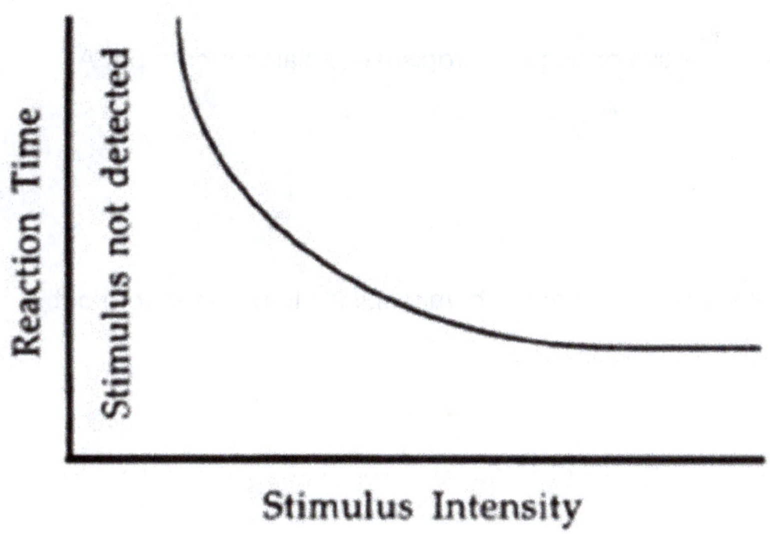

4. What is the difference between sensation (reception) and perception in the nervous system?

5. Hydrotropism is a plant's response to water. To investigate this response, you set up an experiment. In Pot A, a plant is potted in soil. In Pot B, a plant is potted in soil, and the porous clay pot slowly leaks water. The experimental setup is shown in Figure 7.8.

Figure 7.8. An experimental setup to investigate hydrotropism.

6. Describe the responses of the roots in Pots A and B.

   - ❑ What direction do we expect the roots to grow in each pot?

   - ❑ Is this an example of positive or negative tropism? Explain for both pot A and pot B.

   - ❑ How does the plant's response relate to homeostasis? Explain for both pot A and pot B.

Name: ______________________ Lab Time: ____________ Due: ______________

8 Plant Growth Pre-Lab

1. What is a meristem?

2. During what time of year are the cells in wood large in diameter? When are they small in diameter? Why is this?

3. In Activity 8.1, you will construct a scatterplot to see if there is a relationship between two variables. What variables will you be looking at?

4. What is "resource allocation" and how does this relate to homeostasis?

5. What does an R:S ratio measure?

Name: ______________________ Lab Time: ____________ Due: _______________

8 Plant Growth

Homeostasis is responsible for maintaining an organism's internal environment; however, it doesn't exist in a vacuum. Environmental (external) changes will affect homeostasis. Plants are especially affected by external events because, unlike animals, they cannot just migrate somewhere else. Consequently, a plant's ability to regulate homeostasis is directly linked to the factors it's exposed to.

Learning Outcomes

- ❑ Describe the formation of wood and how wood anatomy relates to homeostasis and environmental conditions.
- ❑ Discuss how plant growth is affected by nutrient availability.

Activity 8.1. Wood Analysis

Woody plants leave behind a record of the environment in their wood. *Wood* is formally called **secondary growth** and is due to a tissue called a **vascular cambium** that acts as a meristem. A **meristem** is a location where a plant grows.

Background

In secondary growth, the vascular cambium causes the plant to grow outward (in girth) rather than taller (in height). The tissues it makes are types of *vascular tissue* (remember this?). We call this tissue "wood" when we see it. Figure 8.1 summarizes wood production.

Figure 8.1. A diagram illustrating the production of wood growth.

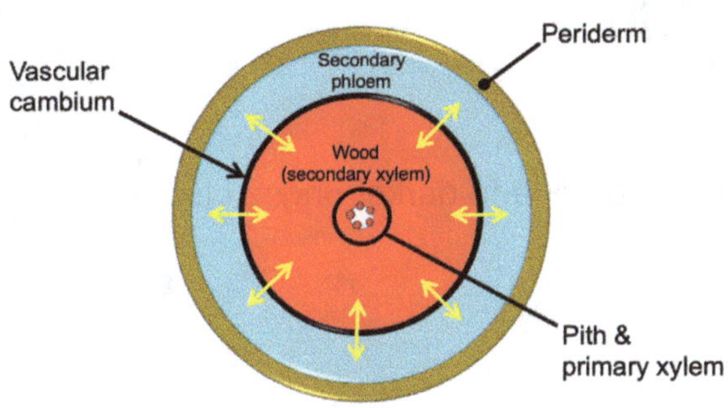

When you look at a cross-section of a tree, you will see a pattern of rings. Each ring is a layer of wood that took a season to grow. This is usually one year, but the tree does not grow at the same rate throughout the year. In our area, it grows very quickly in spring and summer, slowly in autumn, and not at all in winter.

Q1. What factors allow for faster growth in the spring but no growth in the winter?

The light ring is larger because it represents spring and summer growth; the dark ring represents slow autumn growth. When you look at the rings, the contrast between light and dark rings represents one year. Use Figure 8.2 or a provided slide.

Figure 8.2. An example of the contrast between light and dark rings in tree growth.

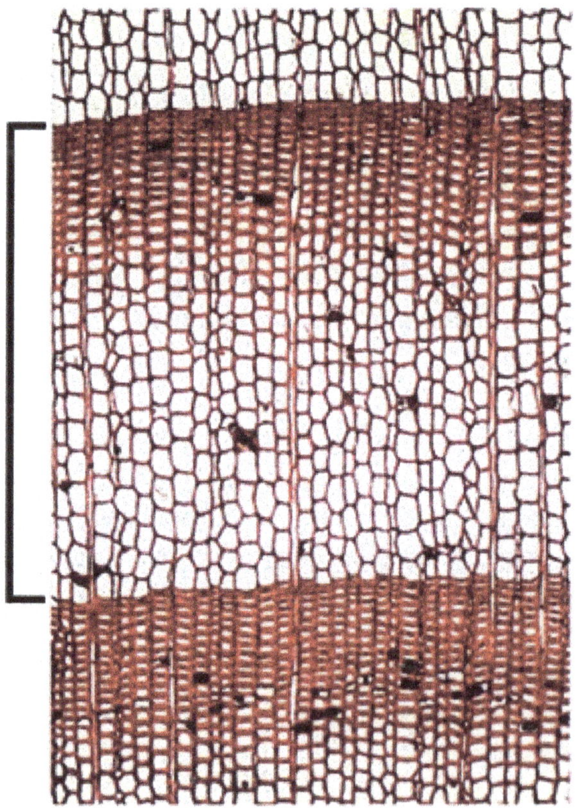

Q2. What differences do you see in the cells of the lighter and darker rings? List two.

Q3. What is the biological reason for your answer above? That is, match the cell anatomy to homeostasis and the factors you listed in the first question.

Procedure A. Does Ring Count Correspond to Diameter?

(adapted from a series of activities from the UK site Science & Plants for Schools https://www.saps.org.uk/)

This experiment will have you test if ring count (age) is related to tree diameter. This relationship is important for extrapolating tree growth in their native environment.

Instructions

1. Use the photos of the tree rounds to count the number of rings from the center of the tree to the bark. The oldest ring is the small one in the middle, and the youngest is the last ring just below the bark (flaky layer on the very outside). The first ring will represent year one (first year of growth).
2. In your group, determine how to be consistent in your counting and measuring.
3. Record your data in Table 8.1 below for pictures A-L.
4. For the tree rounds present in the classroom, have one or two people count the very large round at your lab table and record your data in Table 8.1.
5. Select three smaller tree rounds from the lab counter on the side of the room to count. Add your data for the small rounds to Table 8.1.

Table 8.1. Tree Rings and Diameter

Tree Sample #	Rings counted	Diameter (in cm)	Tree Sample #	Rings counted	Diameter (in cm)
Picture A			Picture I		
Picture B			Picture J		
Picture C			Picture K		
Picture D			Picture L		
Picture E			Large round		
Picture F			Small round 1		
Picture G			Small round 2		
Picture H			Small round 3		

Figure 8.3 is an example of a scatter plot with a best fit line.

Figure 8.3. Example of a scatter plot with a best fit line.

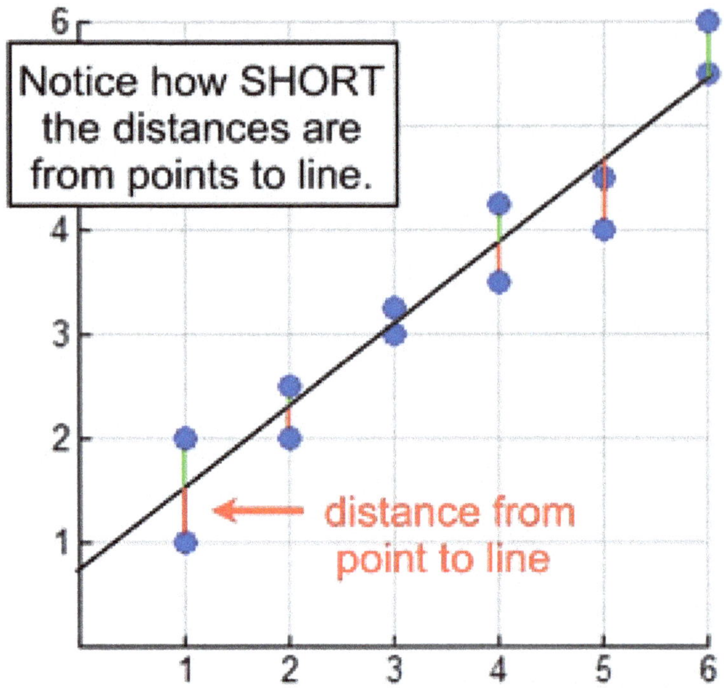

8

6. Make a scatter plot of the data points on Figure 8.4.
7. Place the diameter (cm) on the X-axis and the number of rings (years) on the Y-axis.
8. After plotting your data points, use a ruler to draw a "best fit" line showing the general trend.

DO NOT connect each individual point.

Figure 8.4. Scatterplot: Relationship between the number of rings (years) and a tree's diameter.

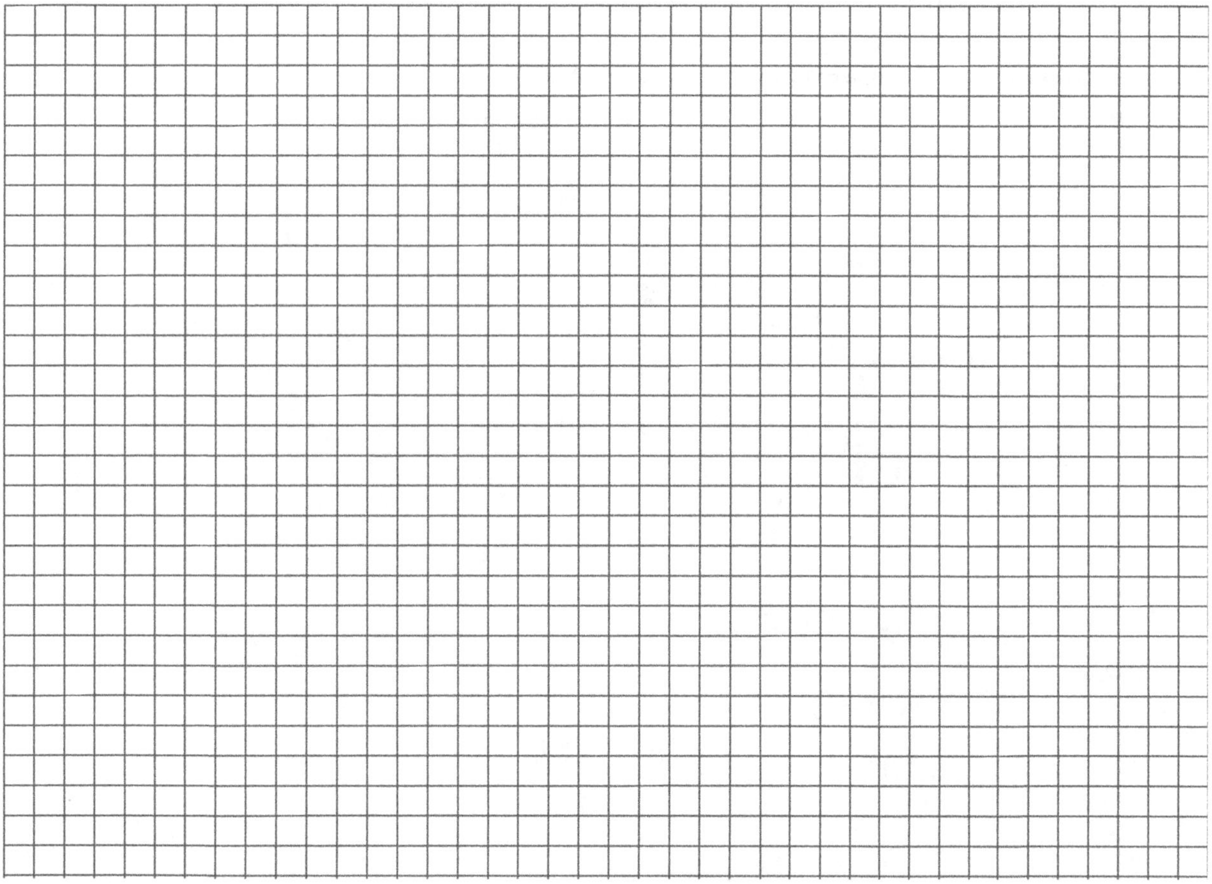

Q4. Does your graph show a direct relationship between age and diameter?

Q5. Do any of your samples lie outside of the "best fit" line–i.e., are outliers? If so, why might that be?

Foresters have a concept called the **species growth factor**, which describes how quickly a tree grows in diameter for its age. Its units are yr/cm (or yr/inch). It is the slope of the line in your graph.

9. Calculate the average slope of your graph.

Slope is rise/run (change in height divided by change in length), which, in this case, is the change in diameter divided by the change in age.

10. Use two X and Y coordinates from your best-fit line, along with the equation in the example in Figure 8.5 to calculate the slope, which is your tree species growth factor.

Figure 8.5. An example of slope calculation using an equation.

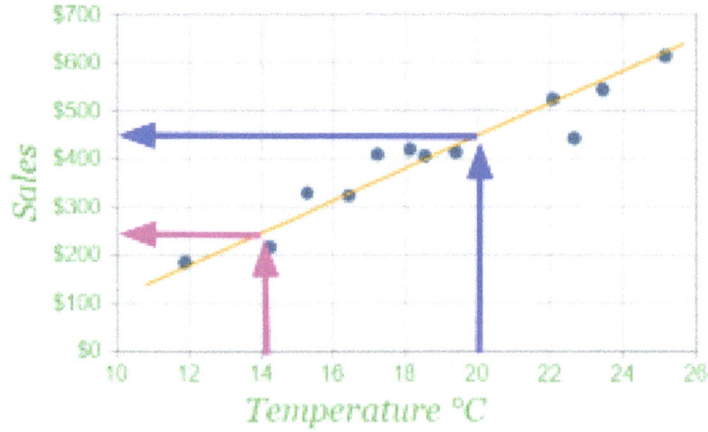

$$\text{Slope} = \frac{\text{rise}}{\text{run}} = \frac{y_2 - y_1}{x_2 - x_1}$$

$$= \frac{450 - 250}{20 - 14}$$

$$= 33.3 \ \$/^{\circ}C$$

Our tree species growth factor is ________________ yr/cm.

Foresters can multiply a tree's growth factor by the diameter at breast height (dbh) to estimate its age without cutting it down, or, conversely, estimate the diameter a tree will reach at a certain age. (fun fact: breast height is set at 4.5 feet off the ground).

Q6. If your trees reached 75 years of age, what would their expected dbh be?

Q7. In the late 1800s near Vancouver, Washington, a Douglas fir was felled having a diameter (dbh) of 16 feet. Using your formula, how old was this tree? (1 foot = 30.5 cm)

Q8. Different tree species, and even the same species in different areas, have different growth factors. For example, a cottonwood's factor is 2.0; an oak's factor is 5.0. Which of these two species (the cottonwood or the oak) would have a larger diameter after ten years of growth? How much larger will its diameter be?

Procedure B. Can Tree Cores Be Aligned to Historical Time?

(adapted from: UCAR Center for Science Education scied.ucar.edu)

Instead of counting rings in a round piece of wood, foresters can use an **increment borer** to take a sample of the wood (a "core") from the tree while it's standing. The core will show rings just like a whole piece of wood. Scan QR Code 8.1 or use the URL (go.chemeketa.edu/increment) to watch the video that shows how an increment borer is used and interpreted:

QR Code 8.1. Increment Borer

This experiment will have you align cores to determine the ages of trees, when they were harvested or sampled, and how climate might have affected the area at those times.

You need the cores to answer the questions from this activity. Make sure all questions are answered before you leave the lab.

Instructions

1. Take the four laminated core samples provided in the lab. Each one represents a different tree from the same region, Pinetown Forest:
    a. **Core #1:** From a living tree sampled in fall 2018.
    b. **Core #2:** From a Christmas tree harvested nearby a few years prior to 2018.
    c. **Core #3:** From a log found near the main trail in Pinetown Forest.
    d. **Core #4:** From a barn beam removed from Pinetown, likely made locally.
2. Count the total rings in each core. Do not include the pith or bark. Record in Table 8.2.
3. Using the known date for Core #1 (taken in 2018), you'll "work backward" to calculate the year the tree was sampled/harvested and the year growth began.

4. Align Core #2 alongside Core #1 until the rings match. This will establish when Core #2's tree existed in relation to Core #1's tree. Calculate as above.
5. Repeat for Core #3 and Core #4, aligning them alongside Cores #1 and #2.
6. Take a picture of your aligned cores so you can keep a record for the questions that follow. You can use a wet-erase marker to write ten-year increments on the cores so you can easily track dates.

Table 8.2. Core Data

Core	Total age (years)	Year cut or cored	Year growth began
#1		2018	
#2			
#3			
#4			

8

Q9. Looking at the ring differences, what year(s) did Pinetown have drought conditions? How did you know? How does this relate to homeostasis?

Q10. During the entire sampling period, did Pinetown have more years of drought or of plentiful rainfall?

Q11. For this activity, we are assuming each tree growing in the Pinetown forest was exposed to the same conditions in the same year. Is this an accurate assumption? Why or why not?

Q12. Where (or what) would a scientist have to core to find data on tree growth from prehistoric times, given that there are no currently alive trees that old?

Activity 8.2. Plant Resource Responses

(adapted from: Johnson, A.D. 2006. Resource Allocation in Plants. Pages 15-30, in Tested Studies for Laboratory Teaching. Proceedings of the 27th Workshop/Conference of the Association for Biology Laboratory Education (ABLE))

Plants can extract only limited resources from their environment, yet they must maintain homeostasis. To do this, plants have to decide where to allocate those nutrients. This experiment will analyze these changes.

Background

Plants require essential resources such as sunlight, nitrogen, and water. When these resources are limited, they affect which species are present in a community. For example, an abandoned field has abundant sunlight but fairly dry soil. How could a plant species survive there? Part of the answer is **resource allocation**: where the plant puts its time and energy given the limitations. This concept is illustrated in Figure 8.6.

Figure 8.6. Locations where plants can allocate energy and nutrient resources.

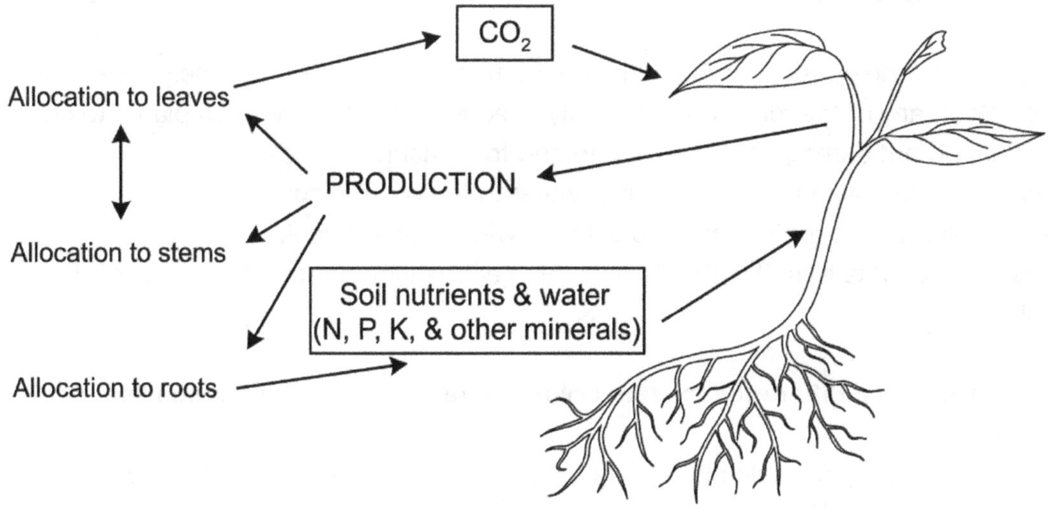

Some plants respond to drought by shifting resources to rapid downward growth of existing roots, while others reallocate their resources to forming a waxy protective cuticle on leaves.

Resource allocation involves trade-offs. Allocating more growth to a stem may increase a plant's height or sugar production but reduce water availability or the plant's ability to remain anchored in poor soil. Conversely, more root growth can lead to increased water supply, but reduced photosynthesis or reproduction.

In this lab, you will measure the root and shoot biomass, then express them as a percentage (the **R:S ratio**). The larger the number, the larger the roots relative to the shoots. The smaller the number, the larger the shoots relative to the roots. An R:S ratio of 1.0 (100%) means the plant is allocating equal resources to both.

Procedure A. Measuring the Root and Shoot Biomass

Your instructor will assign each group one or two species/nutrient combinations to analyze (depending on class size). The number of plants to sample will be:

- ❑ 10 plants if assigned small plants like alfalfa or ryegrass
- ❑ 5 plants if assigned medium plants like oats or radish
- ❑ 1 or 2 plants if assigned very large plants like beans

Your instructor will verify the number of plants to use before you begin.

8

You will harvest and measure 2 single plants, or 2 sets of 5 or 10 plants, for each assigned group.

Instructions

1. Record the species and nutrient concentrations assigned to you in Tables 8.3, 8.4, and 8.5.
2. Carefully pull the correct number of plants from the soil tray. Be extremely careful not to break off any of the root mass. You may need to pull out some extra plants to ensure you do not damage the ones you intend to measure.
3. Gently rinse the plants in a container of water to remove the soil.
4. Gently separate the roots. Pat the plants dry with a paper towel.
5. Separate the roots from the shoots by cutting each plant apart at the "soil line" (the original soil level).

Hint: usually there is a difference in stem color above versus below the soil line.

6. If you are sampling a single plant, weigh the shoot alone, then the root alone. If you are measuring 5 or 10 plants, weigh all the roots together or all the shoots together.
7. Record the total wt.
8. Calculate the weight of a single root or shoot by dividing the total weight by the number of plants in that sample.

9. Record in the individual wt.
10. Calculate the R:S by dividing the root weight (per plant) by the shoot weight (per plant). Average these for both trials. Fill in Table 8.3 with your calculations.

Table 8.3. First Assigned Species

Plant Species: Nutrient Condition:						
	# plants sampled	Roots (g)		Shoots (g)		R:S (as a %) ** (root wt per plant / shoot wt per plant)
		Total wt	per plant	Total wt	per plant	
Trial #1						
Trial #2						
Avg.						

Table 8.4. Second Assigned Species

Plant Species: Nutrient Condition:						
	# plants sampled	Roots (g)		Shoots (g)		R:S (as a %) ** (root wt per plant / shoot wt per plant)
		Total wt	per plant	Total wt	per plant	
Trial #1						
Trial #2						
Avg.						

Table 8.5. Third Assigned Species

Plant Species: Nutrient Condition:						
	# plants sampled	Roots (g)		Shoots (g)		R:S (as a %) ** (root wt per plant / shoot wt per plant)
		Total wt	per plant	Total wt	per plant	
Trial #1						
Trial #2						
Avg.						

**** If R:S = 100%, roots equal shoots. If R:S > 100%, there are more roots than shoots (roots outweigh shoots). If R:S < 100%, fewer roots than shoots (shoots outweigh roots).**

Your instructor will put a table on the board to record class values.

11. Once the entire class has reported their results, copy the table from the board to Table 8.6 below.

Table 8.6. Average Root: Shoot ratios of the three species analyzed during class.

Species of plant	average R:S (%)	
	Low nutrients	High nutrients

Did you see a general pattern in how allocation changes with the amount of nutrient available? If so, what was the pattern, and what is your evidence?

Q13. What factors other than nutrient or light conditions might affect plants' R:S ratio?

8

Q14. Were there any species that did not change their R:S significantly between treatments? If so, what would that indicate about that species' ability to adapt to a different environment?

Q15. Trees can also show changes in R:S. How do you think the R:S of an Oregon ash (a species growing in temperate, extremely wet regions) would compare to that of an alpine fir (a species growing in cold, dry regions)? Why?

Name: ______________________ Lab Time: ____________ Due: ______________

8 Applying What You've Learned

1. Dendroclimatology is a field of study that uses tree rings to reconstruct past climate conditions.

   - ❑ Why can scientists do this? What is a tree ring?

   - ❑ Explain how the growth of a tree provides clues about past climate. What would you observe in a wet climate? a drought? a cold winter? an active fire season? etc.

2. Think about the relative contributions of roots, stems, and leaves to a plant's survival. Which R:S type (high or low) is best suited for the following:

   - ❑ living in the diffuse light of a forest understory? Why?

   - ❑ living on a windy summit near the tree line? Why?

   - ❑ living in a seasonally dry climate like that found in southern California where they typically have a rainy winter followed by a dry summer? Why?

8

Name: ______________________ Lab Time: ____________ Due: ______________

9 Plant Reproduction & Development Pre-Lab

1. What are the two "generations" made in plants' life cycles? Which one is haploid and which is diploid?

2. Where do you look to find the spores of a fern?

3. What is the name for the male part of a flower? What about the female part?

4. How many petals does a monocot have?

5. What is the relationship between a fruit and a seed?

Name: ______________________ Lab Time: ____________ Due: ______________

9 Plant Reproduction & Development

A **life cycle** represents the different stages an organism goes through as it matures and develops. All living things have a life cycle; for example, humans go from baby to child to adult.

In this lab, we will briefly explore the variation in life cycles found within the plant kingdom. We will also investigate the structure and function of flowering plants.

Learning Outcomes

- ❑ Identify sporophyte and gametophyte stages of major plant groups.
- ❑ Describe seeds and fruits and their relationships to plant reproduction.

Background: Alternation of Generation

In animals, meiosis occurs in the diploid **gonads** (testes and ovaries) and produces specialized haploid **gametes** (sperm and eggs). Fusion of these gametes (fertilization or conception) gives rise to a new organism (zygote or offspring).

In plants, by contrast, meiosis produces haploid **spores** that divide by mitosis and grow into a haploid multicellular organism called the **gametophyte generation**. This haploid stage produces gametes (sperm and egg) through the process of mitosis. These **gametes** fuse and form a diploid zygote that undergoes mitosis and grows into a multicellular diploid stage called the **sporophyte generation**. This **alternation of generations**, from haploid gametophyte to diploid sporophyte, is characteristic of all plants.

Q1. Figure 9.1 illustrates a general plant life cycle. Label the diagram in Figure 9.1 by circling the correct word choice.

9

Figure 9.1. The general life cycle of a plant.

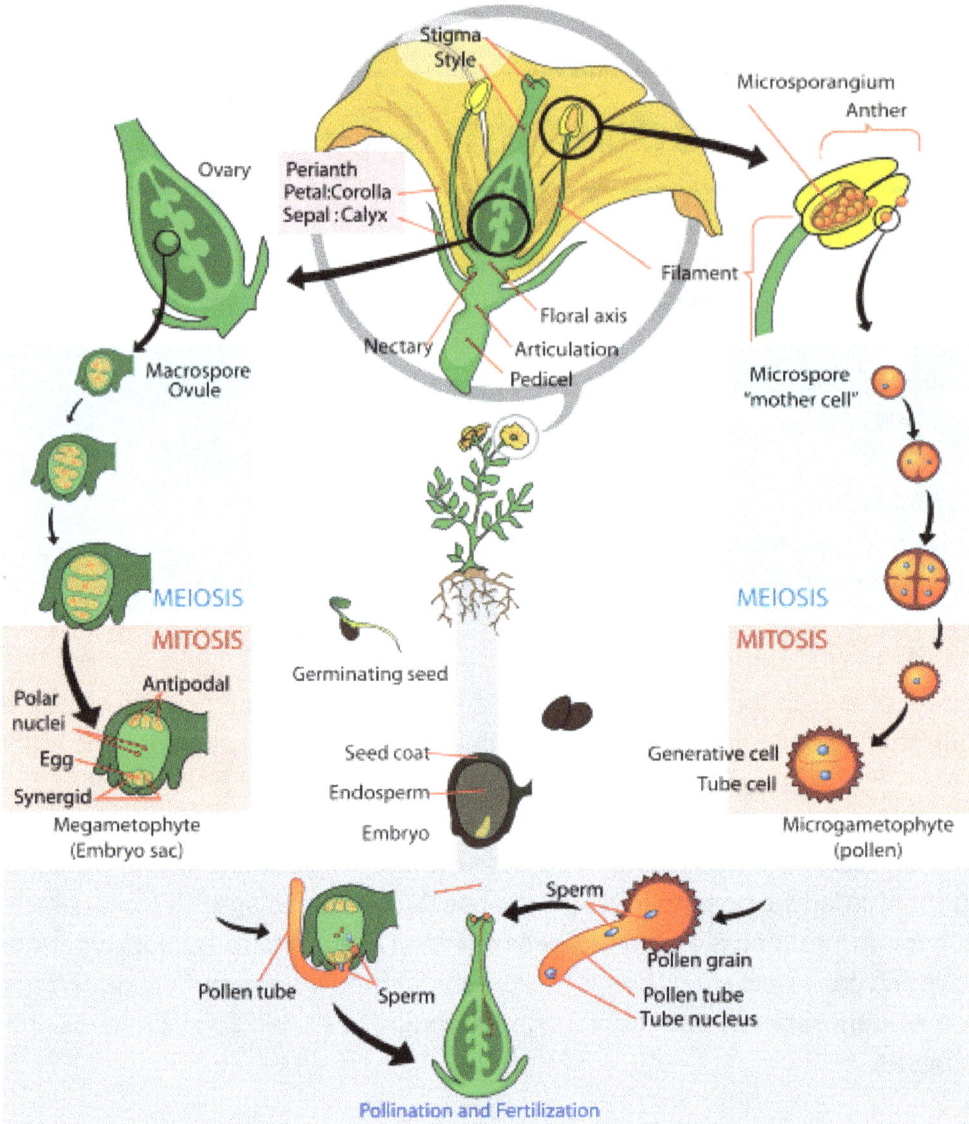

Q2. List one way in which alternation of generation is different from an animal life cycle. In what way is it the same?

Q3. What does it mean if something is diploid? What about haploid?

Activity 9.1. Variation in Plant Life Cycles

Different groups of plants undergo different variations of the plant life cycle and produce different-looking reproductive structures. These structures can be dominant or reduced. In this context, the **dominant stage** represents the larger, more visible part of the cycle. The **reduced stage** is usually less noticeable or even microscopic.

Procedure A. Moss Life Cycle

Examine Figure 9.2, which illustrates the life cycle of a moss (bryophyte).

Figure 9.2. The general life cycle of a moss.

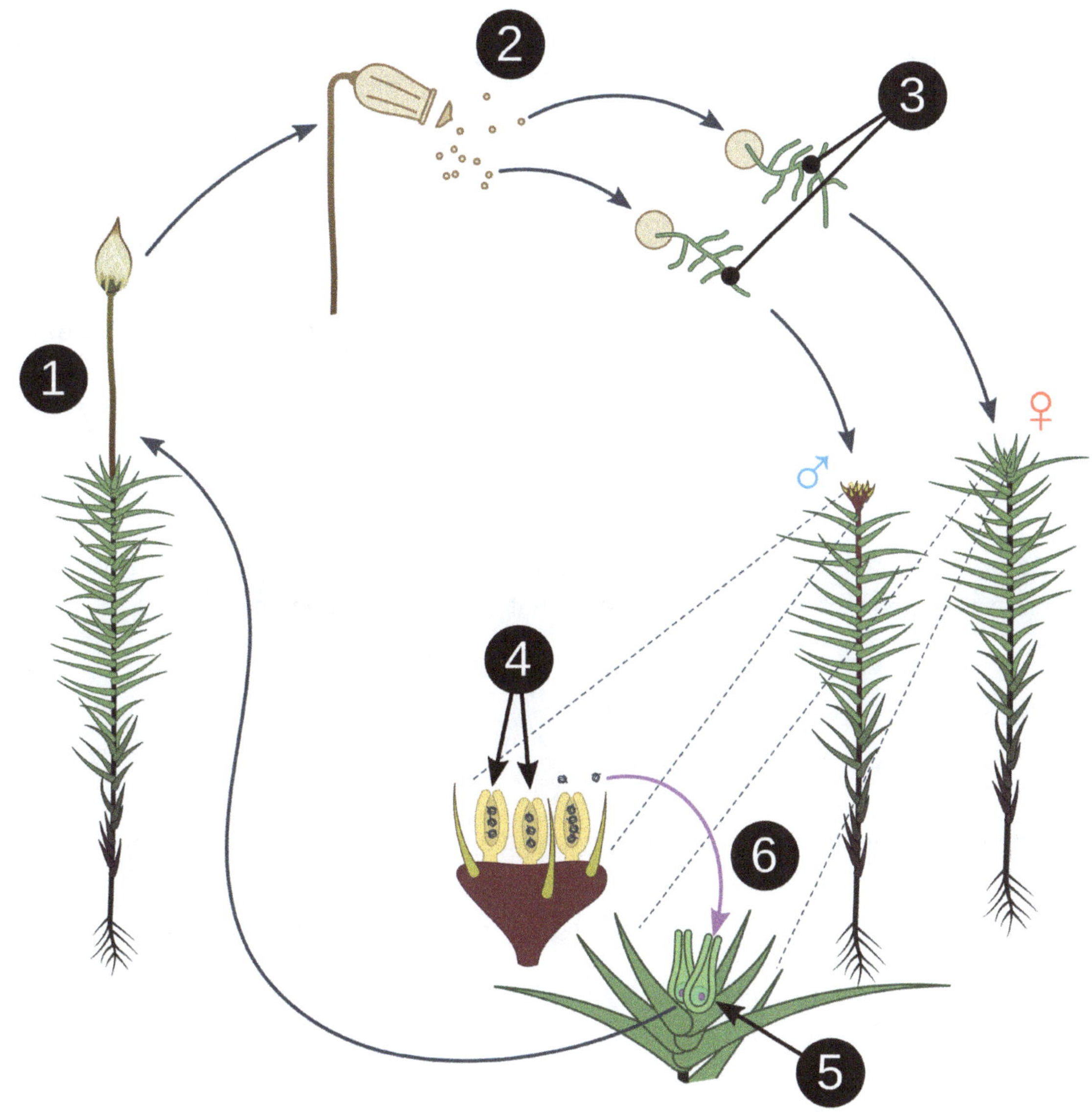

Q4. What number(s) in the diagram best represent the sporophyte stage? How do you know?

Q5. The male and female parts of a moss are on completely separate plants. Given that, how do you think the male gamete (sperm) gets transported to the female gamete (eggs) in a moss?

Q6. Examine the living moss in the lab.

- ❑ What do you notice about the color and location of the sporophyte stage?

- ❑ What do you notice about the color and location of the gametophyte stage?

Q7. Based on your observations of color and location, explain why mosses are said to have a "dependent sporophyte stage." What makes that stage dependent?

Q8. Identify and sketch a moss gametophyte and a moss sporophyte in Figure 9.3.

Figure 9.3. Sketch and identify a moss gametophyte and a moss sporophyte.

Moss gametophyte	Moss sporophyte

Q9. Which life stage (sporophyte or gametophyte) is more dominant in a moss?

Activity 9.2. Fern Life Cycle

A fern's spores are housed in structures called **sori** located under the **fronds**.

Q10. Based upon this statement, are the fronds part of the sporophyte or gametophyte stage?

Q11. Examine the ferns provided in class and sketch one sporophyte in Figure 9.4. Label the sori.

Q12. Examine the slide of fern gametophytes and sketch one gametophyte in Figure 9.4.

Figure 9.4 Sketch and label the gametophytes and sporophytes.

Fern gametophyte	Fern sporophyte

Q13. Which stage of the fern life cycle would you consider dominant?

Q14. The fern sporophyte grows out of the top of the fern gametophyte. However, it does not use the gametophyte as a source of nutrition. Both stages of the fern life cycle are free-living. What do you see that supports this statement?

Q15. The fern gametophyte contains both male and female structures on the same plant. Earlier, the moss had separate male and female plants. What are the advantages and disadvantages of having separate male and female plants versus having both on the same plant?

Activity 9.3. Conifer Life Cycle

Examine Figure 9.5, which depicts the life cycle of a conifer (cone-bearing) plant, a group known as **gymnosperms**.

Figure 9.5. The life cycle of a conifer (cone-bearing) plant.

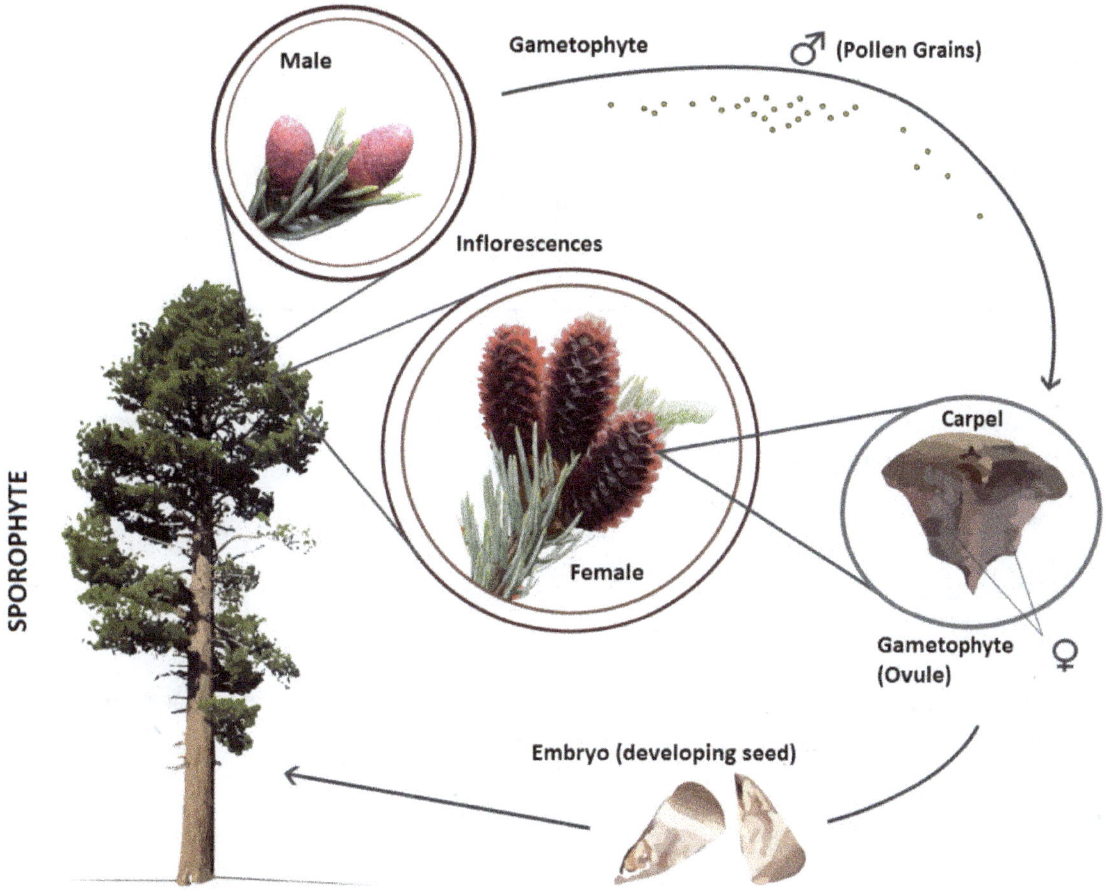

Q16. What stage of the life cycle is dominant–the gametophyte or the sporophyte?

Q17. Examine the cones provided in class and sketch one male (staminate) and one female (ovulate) cone in Figure 9.6.

Figure 9.6. Sketch a staminate and ovulate.

Male cone	Female cone

Q18. What differences do you see between the male and female cones? Why might these differences exist?

Q19. How do you think the conifer's gametes are transported? How does cone anatomy and cone position on the tree relate to this?

Activity 9.4. Flower Structure

As with the conifers in Figure 9.6, flowering plants (**angiosperms**) have a reduced gametophyte buried inside the reproductive organ called a **flower**. The sporophyte stage is dominant and represents the plants that we are familiar with.

Specialized structures within the flower produce spores that develop into the gametophyte. The male gametophyte is called a **pollen grain** and is produced in a structure called an **anther**. The female gametophyte, called the **embryo sac**, is produced within a structure called the **ovule**.

Figure 9.7 shows the main parts of a flower. These parts include:

- ❑ **Sepals**: first whorl at the bottom. These enclose the flower when it's closed.
- ❑ **Petals**: second whorl from the bottom. These are often colorful.
- ❑ **Stamen**: the male part, consisting of an anther on top of a long filament.
- ❑ **Carpel**: the female part, consisting of a stigma and a long style that descends to the round ovary.
- ❑ **Ovules**: located inside the ovary and contain the egg cell.

Q20. Using the descriptions of the main parts of a flower, color in the parts of the flower in Figure 9.7.

Figure 9.7. Color in the main parts of the flower.

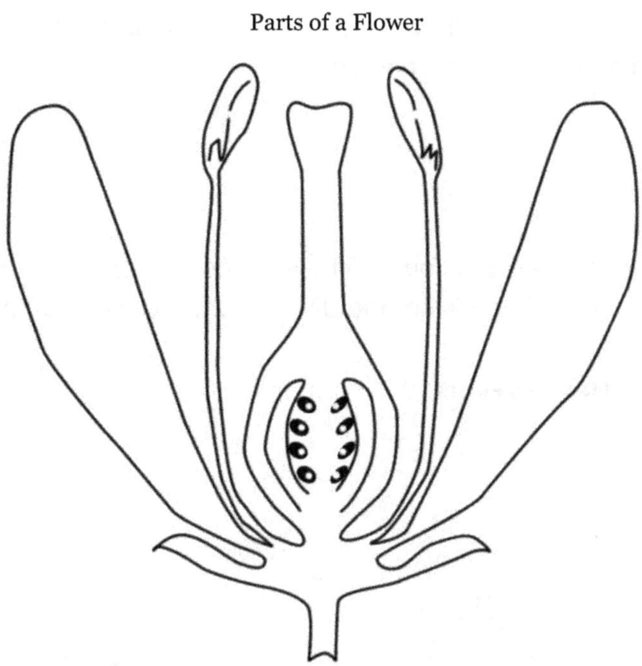

- ❑ Sepals: green
- ❑ Petals: red
- ❑ Stamens: yellow
- ❑ Carpel: brown
- ❑ Ovule: blue

Q21. What part(s) in your diagram above represent the gametophyte stage?

The flower you examined contained both male and female reproductive organs.

Q22. What is an advantage to this arrangement, genetically or ecologically?

Q23. What is a disadvantage to this arrangement?

Q24. Some plants have separate male and female flowers (or even whole plants). What is an advantage and a disadvantage of this arrangement?

Fill in Table 9.1 using the flowers on display in the lab. You may be required to dissect the flower. You can watch a video to assist you using QR Code 9.1 or the URL (go.chemeketa.edu/flower).

QR Code 9.1. Flower Dissection: Reproduction in Flowering Plants

Table 9.1. Flower Dissection

	Plant #1	Plant #2	Plant #3
What color are the sepals?			
How many sepals are there?			
What color are the petals?			
How many petals are there?			
How many stamens on one flower?			
Is the stigma branched? (y/n)			
How many ovules are in the ovary?			

Flowering plants can be divided into two groups called monocots and dicots. A **monocot** generally has flower parts in multiples of three. **Dicots** generally have flower parts in fours or fives.

There are other differences; some we'll discuss later in this lab.

Q25. Based upon the petals and sepals, which of your flowers is a monocot? Which is a dicot?

Q26. Do the numbers of stamens and/or ovules show a pattern between your monocot and dicot flowers? Explain.

Q27. Now look at the stem and leaves of the flowers. What differences do you see between the monocots and dicots?

Activity 9.5. Pollination Biology

To produce offspring, plants must have the male sperm cell, located inside the pollen grain, reach the egg cell inside the ovule. This process is called **pollination**.

The pollen grain contains two **sperm** and a pollen tube nucleus. After pollination, **the tube nucleus** grows a pollen tube towards the ovules, followed by the sperm. Once at the ovule, one sperm fuses with the egg to form the **embryo**, and the other sperm fuses with two **polar nuclei** to form the **triploid endosperm**. This is called **double fertilization**.

These plant processes are depicted in Figure 9.8.

Figure 9.8. The pollination process.

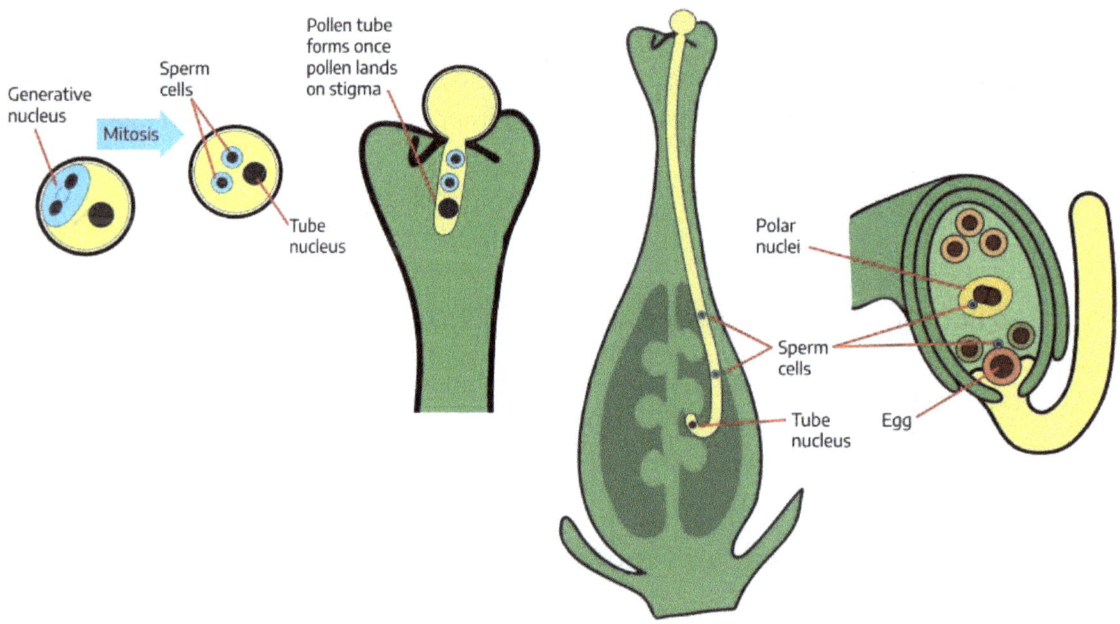

Procedure A. Studying Pollen

Instructions

1. Examine a slide of "Pollen Tubes."
2. Sketch a germinating pollen grain in Figure 9.9.
3. The pollen tube nucleus is normally located at the very tip of the pollen tube, and the sperm are located further up the tube.

You may have to move your fine focus knob up and down quite a bit to find all three nuclei.

4. Label the pollen grain, the pollen tube nucleus, and the two sperm in Figure 9.9.

Figure 9.9. Sketch and label a germinating pollen grain.

Pollen tubes

Q28. In double fertilization, which two structures are fertilized, and what does each produce?

Q29. What is the benefit to the plant of producing endosperm at the same time as an embryo?

Q30. We defined pollination in the introduction to Activity 9.5. What is cross-pollination and why is it important? What feature(s) on flowers increase the probability of cross-pollination?

Getting pollen from one plant to another is critical for reproduction, and plants use different strategies to get the job done. Many are wind-pollinated. Others attract animals–mostly insects and a few types of birds–to move pollen. Some plants offer **nectar** as a sweet reward for pollinators. While any one pollinator may visit many species of flowers, and any one flower may attract many different pollinators, we can get a general idea of what might pollinate a plant by the flower's features. Table 9.2 shows some general trends.

Table 9.2. Flowers That Attract Pollinators

Pollinator type	Flower features
Honey bees	Open, platform-shaped flowers that offer both nectar and pollen. Purple, violet, and blue flowers that show ultraviolet lines on petals.
Butterflies	Narrow tubes that only the butterfly's long tongue can reach into. Colors similar to those for honey bees.
Hummingbirds	Flowers may hang down, making it easier for the hummingbird to approach. Preference is bright red flowers.
Moths	Night-blooming, bright white, and often strongly scented. The nectar that the moths seek may be at the bottom of a long, narrow tube.

5. Select three different flowers (these could be the flowers you dissected, or others). Make a prediction about the possible pollinator in Table 9.3.

Table 9.3. Production of Plants and Their Pollinators

	Flower 1	Flower 2	Flower 3
Probably pollinated by . . .?			
My reasoning:			

Q31. A scientist discovered a plant pollinated by bats. What characteristics of this flower would you expect to see?

Q32. If ants pollinated a plant, would you expect the flowers to be large or small? Why?

Activity 9.6. Seeds and Germination

After pollination, the ovule changes. The contents of the endosperm become embryonic leaves known as **cotyledons**. The cotyledons are the thick, tough leaves that first appear when you plant a bean seedling. The terms monocot and dicot from earlier in the lab refer to these cotyledons. The ovule develops a hardened outside layer called the **seed coat**. This makes the ovule into a **seed**. The ovary changes into either a fleshy or hard structure, which is now called a **fruit**. Thus, a fruit contains the seeds and the seeds contain an embryo with endosperm. "A fruit is a baby in a box with its lunch."

Procedure A. Examining the Germination Process

Instructions

1. Obtain a bean seed that has been soaked in water for 24 hours and a bean seed that has been germinated for 2 weeks (a seedling).
2. Sketch your seedling and your soaked seed in Figure 9.10.
3. Label the shoot, root, and cotyledons of the germinated seed and the embryo and cotyledons of the soaked seed in Figure 9.10.

Figure 9.10. Sketch and label a soaked bean and a germinated bean.

Soaked seed	Germinated seedling

Energy in plants is stored in the chemical bonds of sugars. When the plant has excess sugars, it converts them to starch, a molecule for long-term sugar storage. An iodine solution will react with starch to create a purple-black color. Sugars do not react with iodine solution.

Q33. Where in a seed or a germinated seedling do you think we might find starch?

Q34. If you stain the seedling and the soaked seed with iodine solution, which would you expect to contain starch? Why?

4. Add a few drops of iodine solution to the appropriate locations on your soaked and germinated seeds.

Q35. Did either your soaked seed or your seedling stain positive for starch? Which one?

Q36. Based on this experiment, what role does starch play in a seed?

Q37. How does the production of seeds benefit flowering plants?

Activity 9.7. Fruits

The purpose of the fruit is **seed dispersal**, moving the seeds away from the parent plant. In Activity 9.7, we will examine several different types of fruit. As you look at each fruit, consider the role the environment played in the evolution of its form. Examples of several types of fruit are provided in class. Some are called **fleshy fruits** and others are called **dry fruits**.

Q38. Compare at least three fleshy fruits and three dry fruits and fill Table 9.4.

Table 9.4. Comparing Fleshy and Dry Fruits

Fruit name	Fleshy or dry	Description of fruit	How dispersed? How do you know?

9

Q39. List a few other ways fruit may be dispersed besides the fruits/methods listed in the table above.

Development of a Fruit

Fruits are the result of the development of various floral parts. In the example of a peach, the inner lining of the ovary becomes what we know as the pit, the outer lining of the ovary becomes the fuzzy skin, and the middle tissue of the ovary becomes the sweet fleshy tissue.

Instructions

1. Sketch and label the seed, seed coat, and fleshy parts of the fruit in Figure 9.11.
2. Sketch and label the seed and seed coat of the dried fruit in Figure 9.11.

Figure 9.11. Sketch and label the seed and seed coat of fleshy and dried fruits.

Fleshy fruit	Dried fruit

Q40. Which of these are a fruit: cucumber, lettuce, carrot, celery, peas, tomato, pumpkin, peanut? Explain your answer.

Q41. Some people consider potatoes a fruit, but it isn't. Why isn't it?

Q42. What part of a flower creates a banana peel? How do you know?

9

Name: ______________________ Lab Time: ____________ Due: ______________

9 Applying What You've Learned

1. What happened to the relative size of the sporophyte and gametophyte phases of the plant life cycle as plants evolved within drier habitats?

2. Compare spores with seeds and explain why seeds are advantageous to land plants.

3. Corn (like all grasses) is a flowering plant.
   - ❑ What would be the fruit and seeds of a corn plant?
   - ❑ Farmers can "detassel" corn plants to prevent those plants from producing pollen. Detasseling is the process of cutting off the top part of the plant (the tassel). Based upon this, what reproductive structure(s) must the tassel contain?
   - ❑ Detasseling is usually done to force cross-pollination. How would this be effective?

4. In most conifers, male cones are usually on lower branches, while female cones are on higher branches. Also, female cones are generally larger and more conspicuous than male cones. Explain both of these points in light of a conifer's life cycle.

9

Acknowledgments

Lab 1

Figure 1.1. " Different types of epithelial tissue" by Wikipedia. Public Domain. https://en.wikipedia.org/wiki/Epithelium#/media/File:Illu_epithelium.jpg.

Figure 1.2. "Sketch and label the tissues present in the seal surgery photos," by The Marine Mammal Anatomy and Pathology Library. https://mmapl.ucsc.edu/normal-anatomy-sea-lion/integument-and-pelage.

Figure 1.3. "The average of blubber thickness over several months," by Andre Hermansson. https://www.slideshare.net/slideshow/blubber-thickness-variation-in-grey-harbour-ringed-seal-skorrigerade-energiberkningarslutgiltig/59153994#23

Figure 1.4. "The average blubber thickness over several years," by Andre Hermansson. https://www.slideshare.net/slideshow/blubber-thickness-variation-in-grey-harbour-ringed-sealskorrigerade-energiberkningarslutgiltig/59153994#23.

Figure 1.5. "Bird skeleton with keel bone in blue," by Toony. Wikipedia. Public Domain. https://en.wikipedia.org/wiki/Keel_%28bird_anatomy%29#/media/File:Br%C3%A9chet.png.

Figure 1.6. "Lateral view of a frog, stained for bone (red) and cartilage (blue)," by the University of Chicago Department of Organismal Biology & Anatomy. https://oba.bsd.uchicago.edu/news/new-discovery-shubin-lab-frogs-give-new-insights-vertebrate-novelties.

Lab 2

Figure 2.1. "105 Negative Feedback Loops," by OpenStax. Wikimedia. Public Domain. https://commons.wikimedia.org/wiki/File:105_Negative_Feedback_Loops.jpg.

Figure 2.2. "An example of positive and negative feedback," by Flexbooks 2.0. https://flexbooks.ck12.org/cbook/ck-12-college-human-biology-flexbook-2.0/section/9.7/primary/lesson/homeostasis-and-feedback-chumbio/.

Figure 2.3. "The Oak (Marshall Ward)" by H. Marshall Ward. Wikimedia. Public Domain. https://commons.wikimedia.org/wiki/File:The_Oak_(Marshall_Ward)_Fig_23.jpg

Lab 3

Figure 3.1. "Leaf Anatomy" by H. McKenna. Wikimedia. Public Domain. https://commons.wikimedia.org/wiki/File:Leaf_anatomy.svg.

Figure 3.2. "Investigating the behaviour of leaf discs," by Science & Plants for Schools. https://assets.ctfassets.net/pc40tpn1u6ef/file-15566/9d04f70bc2fe796b2c751322b308ee53/10327-SAPS_Sheet_3_-_Leaf_Discs.pdf.

Figure 3.4a. "Schematic illustration of (A) an absorption spectrum of chlorophyll and (B) optimum light intensity for photo-induced cell aggregation (OLI)," by Ryuhei Nakamura. https://www.researchgate.net/figure/Schematic-illustration-of-A-an-absorption-spectrum-of-chlorophyll-and-B-optimum-light_fig3_262069512.

Figure 3.4b. "The Light-Dependent Reactions of Photosynthesis" by Biology 2e. Original image: NASA. https://opened.cuny.edu/courseware/lesson/645/student/?section=3

Figure 3.6. "Investigating photosynthesis: Testing a leaf for starch," by BBC. https://www.bbc.co.uk/bitesize/topics/z7c72v4/articles/zdxkcmn?trk=public_post_comment-text#znchtrd

Lab 4

Figure 4.3. "Staining reveals different cell types in this light micrograph of a wheat (Triticum) root cross section," by Matt Russell. Biology 2e. OpenEd CUNY. https://opened.cuny.edu/courseware/lesson/763/student-old/?task=3#fig-ch30_03_03.

Figure 4.5. "An example of transpiration in a celery stalk," by MrsKRamos. Slideshare.net. https://www.slideshare.net/slideshow/transpiration-biology-prac-format/29162181#5.

Figure 4.6. "Cross-section view of a celery piece under a microscope, magnified 8x," by biologybnapier. The Biology Bar. https://www.teacherspayteachers.com/store/the-biology-bar.

Lab 5

Figure 5.1. "Segmental anatomy of an earthworm. This schematic drawing shows the basic anatomy of annelids in a cross-sectional view," by Biology 2e. OpenSTAX. https://openstax.org/books/biology-2e/pages/28-4-superphylum-lophotrochozoa-mollusks-and-annelids#fig-ch28_03_17.

Lab 6

Figure 6.1. "Breathing cycle in insects, humans, and birds," by Taeah Kim. University of Toronto Mississauga. https://ecampusontario.pressbooks.pub/animalphysiology/chapter/2-2/.

Lab 7

Figure 7.3. "Anatomy and physiology of animal Relation BTW sensory, relay & motor neurons," by Ruth Lawson. Otago Polytechnic. Wikimedia. Public Domain. https://commons.wikimedia.org/wiki/File:Anatomy_and_physiology_of_animals_Relation_btw_sensory,_relay_%26_motor_neurons.jpg.

Figure 7.4. "Simple Neuromuscular Reflexes," by Vernier. https://www.vernier.com/experiment/hsb-hp-4_simple-neuromuscular-reflexes/?cn-reloaded=1.

Figure 7.7. "The proposed relation between stimulus intensity and reaction time," by Apoorvagiri D. ResearchGate.NET. https://www.researchgate.net/figure/The-proposed-relation-between-stimulus-intensity-and-reaction-time_fig2_236984422.

Figure 7.8. "An experimental setup to investigate hydrotropism," by Board of School Education Haryana and Educational Initiatives. https://bseh.org.in/home.

Lab 8

Figure 8.1. "Relative arrangement of secondary tissues in a mature woody stem," by Digital Atlas of Ancient Life. https://www.digitalatlasofancientlife.org/learn/embryophytes/tracheophytes/.

Figure 8.2. "Microscope cross-sections of fossil redwood (left) and modern coastal redwood (right) demonstrating how they look nearly the same. The black brackets denote a single growth ring," by NPS / Wheeler. National Park Service. https://www.nps.gov/flfo/learn/nature/virtual-learning-from-tree-rings.htm.

Figure 8.3. "Example of a scatter plot with a best fit line," by Math Bits Notebook. https://docs.google.com/spreadsheets/d/12puS5orQYECZr_o5_XjeKAqJeexrzyNH6dbxf_hhmow/edit?gid=0#gid=0.

Figure 8.6. "Locations where plants can allocate energy and nutrient resources," by Association for Biology Laboratory Education. https://www.ableweb.org/.

Lab 9

Figure 9.1. "Angiosperm life cycle diagram," by LadyofHats Mariana Ruiz. Wikimedia Commons. Public Domain. https://commons.wikimedia.org/wiki/File:Angiosperm_life_cycle_diagram-en.svg.

Figure 9.2. "Moss life cycle," by Tomáš Kebert. Wikimedia Commons. Public Domain. https://commons.wikimedia.org/wiki/File:Moss_life_cycle.svg.

Figure 9.5. “Gymnosperm life cycle,” by Jhodlof (tree photograph), JJ Harrison (female cones photograph), Beentree (male cones photograph), MPF (ovuliferous scale and seeds photograph), RoRo (final life cycle diagram). Wikimedia Commons. Public Domain. https://commons.wikimedia.org/wiki/File:Gymnosperm_life_cycle_%28en%29.png.

Figure 9.6. “Parts of a Flower,” by Vicki Olds. LoveToKnow. https://www.lovetoknow.com/home/garden/parts-flower.

Figure 9.8. “Parts of a Flower,” by OpenStax Biology. University of Michigan. https://open.lib.umn.edu/horticulture/chapter/14-1-gametogenesis/.

www.ingramcontent.com/pod-product-compliance
Lightning Source LLC
Chambersburg PA
CBHW081245140926
51296CB00002B/1
* 9 7 8 1 9 5 5 4 9 9 3 3 0 *